RECENT ADVANCES IN
ANIMAL NUTRITION—1995

Cover design

Mr Loveden's Berkshire Hog. From Mavor (1809), Agricultural Reports for England and Wales, drawn up for the Board of Agriculture—Berkshire. For further details, see Wiseman (1986) A History of the British Pig, London: Duckworth.

Recent Advances in Animal Nutrition

1995

P.C. Garnsworthy, PhD

D.J.A. Cole, PhD
University of Nottingham

NOTTINGHAM
University Press

Nottingham University Press
Manor Farm, Main Street, Thrumpton
Nottingham NG11 0AY, United Kingdom

NOTTINGHAM

First published 1995

British Library Cataloguing in Publication Data
Recent Advances in Animal Nutrition—1995:
University of Nottingham Feed Manufacturers
Conference (29th, 1995, Nottingham)
I. Garnsworthy, Philip C. II. Cole, D.J.A.

ISBN 1-897676-026

Typeset by Create Publishing Services Ltd, Bath, Avon.
Printed and bound by Redwood Books, Trowbridge, Wiltshire

PREFACE

The 29th Feed Manufacturers Conference was held at the University of Nottingham in January 1995 and this book represents the proceedings of that conference. The topics covered were chosen by the committee to provide answers to problems currently facing the animal feed industry or to stimulate constructive thoughts and discussion about future strategies.

The first chapter considers the effects of carbohydrates on rumen fermentation and the extent to which dairy cow performance is likely to be modified as a result. The second chapter describes the advances made in the use of metabolic profiles for dairy cows and how these can be used to monitor their nutritional status. The third paper draws on the experience of our US colleagues in feeding cows with high genetic merit and concludes that dry matter intake is the major limitation to production. In the UK, grass silage is the predominant forage for dairy cows and new ways of predicting intake were discussed in the fourth chapter. Unfortunately, the authors were not prepared to release any of their equations.

In the human food industry there has been a lot of interest in the effects of manufacturing on food properties. For this reason, a food scientist was asked to write the next chapter and extrapolate his knowledge to the animal feed industry. The sixth chapter contains the annual update on legislation relevant to the feed compounder.

Pet food represents an important market for feed manufacturers. Sensory and experiential factors are far more important in the design of foods for domestic dogs and cats than for farmed species and these were reviewed in chapter seven. In the poultry disease industry interest has moved from maximum growth rate of broilers to growth of carcass components. Chapter 8 describes some new models which may assist with the prediction of response to nutrition in these components.

The next two chapters describe recent research on novel feeding systems for pigs. The first looks at alternative, or non-conventional, feeds and how to predict their energy content. The second considers the use of choice-feeding systems for pigs. The final two chapters are concerned with breeding pigs and discuss new findings on the influence of vitamins on reproduction and new ways to calculate their amino acid requirements.

We would like to thank the authors for their valuable contributions to the conference and these proceedings. We would also like to acknowledge the assistance given by Trouw Nutrition towards the running costs of the conference.

P.C. Garnsworthy

D.J.A. Cole

CONTENTS

I

Ruminant Nutrition

1

THE IMPORTANCE OF RATE OF RUMINAL FERMENTATION OF ENERGY SOURCES IN DIETS FOR DAIRY COWS

D.G. CHAMBERLAIN and J-J. CHOUNG
Hannah Research Institute, Ayr, KA6 5HL, UK

Introduction

Much of the current interest in the rate of ruminal fermentation of carbohydrates stems from its relevance to the Metabolizable Protein (MP) system and related systems of protein rationing (Newbold, 1994; Beever and Cottrill, 1994). Taking a longer-term view, knowledge of the rate of fermentation of carbohydrates and the molar composition of the mixture of volatile fatty acids (VFA) produced are essential ingredients of any rationing system that attempts to predict the effects of nutrient supply on the yields of milk constituents (Newbold, 1994), but there is a more immediate interest.

Synchronizing the availability of energy and nitrogen in the rumen is seen as offering considerable potential to enhance the output of microbial protein from the rumen in certain dietary circumstances, many of which are not uncommon in practice. This idea is attractive to the feed compounder, who sees the potential to manipulate rate of fermentation by varying the sources and types of carbohydrate in formulations, thereby matching the rates at which energy and nitrogen from the basal forage and the compound become available to the microbial population.

Part of this paper will be concerned with an examination of the extent to which microbial protein synthesis can be influenced by changes in synchrony of energy and nitrogen release in the rumen, but it is important to note that altering the rate of fermentation is likely to have a number of consequences both for the microbes and the host; some of these are listed in Figure 1.1. From the outset then, we should recognize that there are likely to be limits to our ability to manipulate rate of fermentation without incurring penalties such as reduced feed intake, changes in partition of

Rate of ruminal fermentation of CHO

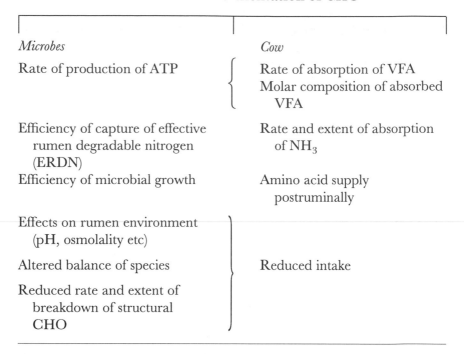

Microbes	*Cow*
Rate of production of ATP	Rate of absorption of VFA Molar composition of absorbed VFA
Efficiency of capture of effective rumen degradable nitrogen (ERDN)	Rate and extent of absorption of NH₃
Efficiency of microbial growth	Amino acid supply postruminally
Effects on rumen environment (pH, osmolality etc) Altered balance of species Reduced rate and extent of breakdown of structural CHO	Reduced intake

Figure 1.1 Some possible effects of rate of fermentation of CHO on ruminal microbes and the host.

nutrient use between the udder and other body tissues and changes in milk composition.

Synchrony of energy and nitrogen release in the rumen

The basic assumption is that a lack of synchrony between the rates at which energy and nitrogen become available to the microbes will lead to a reduced efficiency of microbial capture of nitrogen, and ATP production from fermentation of dietary carbohydrate being inefficiently used for microbial growth. To examine the basis of this assumption in more detail, we may consider the extreme form of asynchrony shown very simply in Figure 1.2.

A simplistic interpretation is to conclude that the peak in the availability of nitrogen (very largely, ammonia) leads to considerable absorption of ammonia from the rumen because there is little energy available to support microbial incorporation of ammonia (and amino acids and peptides). Using similar reasoning, when fermentation of dietary carbohydrate reaches its peak, the ruminal supply of available nitrogen will be markedly deficient, leading to an 'uncoupling' of ATP production and microbial protein synthesis, such that fermentation occurs largely without the growth of microbial cells. It is clear that the soundness of this interpretation rests heavily on our knowledge of factors controlling the absorption of ammonia and the recycling of urea to the rumen, and on our knowledge of the microbial utilization of ATP generated during fermentation.

ABSORPTION OF AMMONIA AND RECYCLING OF ENDOGENOUS NITROGEN TO THE RUMEN

It is generally assumed that the rate of absorption of un-ionized ammonia is much more rapid than that of the ammonium ion. It then follows that the rate of absorption is heavily dependent on ruminal pH: at pH values less than 7, there are only very small proportions of un-ionized ammonia present. Indeed, from an examination of the evidence available, Smith (1975) concluded that, at ruminal pH values less than 6.5, there was little evidence of appreciable absorption of ammonia from the rumen; this did not, of course, preclude absorption of ammonia from the postruminal gut, especially the omasum, following outflow of ammonia in digesta. The overall conclusion was that, under normal feeding conditions, ammonia is only slowly absorbed from the rumen, the implication being that the synchronization of ammonia and energy release in the rumen need not be anywhere near as precise as may sometimes be imagined.

Moreover, if ammonia should be in short supply during the feeding cycle,

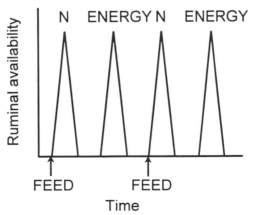

Figure 1.2 A simple diagram to show an extreme form of asynchrony between the ruminal availability of energy and nitrogen.

there is scope to enhance its concentration by way of recycling of urea in saliva or by entry across the ruminal epithelium. It is important to note that an increased rate of fermentation, resulting from the inclusion of readily fermentable carbohydrate in the diet, leads to an increased entry of urea into the rumen as a result of increased permeability of the ruminal epithelium (see Obara, Dellow and Nolan, 1991).

MICROBIAL UTILIZATION OF ATP GENERATED DURING FERMENTATION

Put simply, ATP generated during fermentation is used by the microbes essentially for two purposes: maintenance of cellular function and synthesis of cell constituents. However, the use of ATP for the synthesis of cell material is markedly influenced by the composition of the culture medium; cell composition can vary widely, particularly with respect to the content of storage polysaccharides (see Russell and Wallace, 1988). When the supply of fermentable carbohydrate is in relative excess of the supply of available nitrogen, rumen bacteria use the ATP produced in fermentation to synthesize storage polysaccharide, which can account for as much as 75% of cell dry matter (DM) (Stewart, Paniagua, Dinsdale, Cheng and Garrow, 1981). The energy cost of polysaccharide synthesis is only about one-third of that required for the synthesis of protein (Stouthamer, 1973) and, furthermore, the stored polysaccharide can be mobilized and the glucose metabolized to yield ATP for protein synthesis later in the feeding cycle when adequate supplies of usable nitrogen become available. Of course, this process is somewhat less efficient than the immediate use of ATP for protein synthesis, but these findings again emphasize the ability of ruminal bacteria to cope effectively with fluctuating supplies of energy and nitrogen.

We would be led to believe from the foregoing that, in general, the response of microbial protein synthesis to changes in the pattern of fermentation of carbohydrate and the release of ammonia would be small. Indeed, this view is supported by the results of experiments *in vitro*, even when the degree of mismatching between energy and nitrogen supply patterns was severe (Henning, Steyn and Meissner, 1991; Newbold and Rust, 1992), and by results of experiments *in vivo*. Changes in the frequency of intraruminal dosing with urea failed to influence the utilization of nitrogen in growing lambs (Knight and Owens, 1973; Streeter, Little, Mitchell and Scott, 1973). Again, spreading intraruminal doses of urea and starch or glucose to achieve a better matching of the rates of energy and ammonia release did not influence microbial capture of ammonia (Salter, Smith and Hewitt, 1983). Synchronizing rate of fermentation and rate of ammonia release would be expected to be beneficial only in conditions of extensive absorption of ammonia, such as might occur if ruminal pH values were high (say > 6.8) for significant periods of the feeding cycle. Whether this was the case in the experiment of Meggison, McMeniman and Armstrong (1979) which showed beneficial effects on the microbial capture of ammonia when urea was administered continuously as opposed to twice daily, is not known.

The general conclusion that a close matching of energy and nitrogen release will bring only small, if any, benefit in most practical conditions holds only if microbial growth is limited by the supplies of fermentable carbohydrate and ammonia. We must consider the possibility that other nitrogenous compounds may be limiting microbial protein synthesis, and that the supply of these other nutrients may be affected by changes in feeding strategies. Although the supply of minerals and vitamins can influence bacterial growth rates in the rumen (see Mackie and Therion, 1984), this subject will not be considered further, and attention will be focused on the role of amino acids and peptides as sources of nitrogen for microbial protein synthesis.

AMINO ACIDS AND PEPTIDES AS SOURCES OF NITROGEN FOR MICROBIAL PROTEIN

In normal circumstances, ammonia is the most abundant nitrogen compound available for microbial growth; indeed, most cellulolytic organisms have an absolute requirement for ammonia (Hespell, 1984). On the other hand, starch-digesting bacteria have been shown to obtain 66% of their nitrogen from amino acids and peptides and only 34% from ammonia (Russell, Sniffen and Van Soest, 1983). The implication of findings such as these is that, although cellulolytic bacteria may need only small amounts of amino acids in ruminal fluid, provided their requirements for branched-chain amino acids are met (Bryant 1973), starch-digesting bacteria may need significant amounts of amino acids and peptides if they are to achieve their maximum growth rates, an idea developed by Russell, O'Connor, Fox, Van Soest and Sniffen, (1992).

Amino acids and peptides accumulate in ruminal fluid in the early post-feeding period, the extent of accumulation depending on the degradability of the dietary protein. Unless highly soluble proteins like casein are fed, the concentrations of free amino acids remain low, but concentrations of peptide can reach values in excess of 200 mg N/l just after feeding, rapidly declining to < 25 mg N/l after 2 hours (Broderick, Wallace and Ørskov, 1991).

Hence, the suggestion (Hespell and Bryant, 1979) that bacterial growth may be limited by deficiencies of amino acid and peptide at times during the feeding cycle should be seen as particularly relevant when the diet is rich in starchy concentrates, and the ruminal population contains a high proportion of amylolytic bacteria. It is interesting to note that, in continuous culture studies with diets rich in starch, microbial protein output was linearly related to the intake of rumen-degradable protein up to levels in excess of 20% of DM (Hoover and Stokes, 1991).

From the above, it might be expected that, particularly with diets rich in readily fermentable carbohydrate (RFC), changes in the pattern of feeding, either of the whole diet or of protein or carbohydrate ingredients, would result in improved rates of microbial protein synthesis. To what extent such benefits have been shown to occur in practice is examined below.

EFFECTS OF CHANGES IN FEEDING PATTERN ON MICROBIAL PROTEIN SYNTHESIS

We must recognize that interpreting the results of experiments in which feeding pattern has been altered is not always straightforward, and this is especially true when the diet is rich in RFC. If the frequency of feeding of the whole diet or of the RFC component has been altered, often there are pronounced effects on ruminal pH, the molar composition of VFA etc, which may influence microbial growth in their own right. Probably, the experimental approach that offers the clearest interpretation is one in which the frequency of feeding of the protein component of the diet is altered, whilst all other components of the diet are maintained constant. We are aware of only one published study that comes close to meeting these criteria. Robinson and McQueen (1994) examined the effects on the performance of dairy cows of feeding two protein supplements either twice or five times daily. There were no effects on milk production, but no conclusions can be drawn because altering the frequency of feeding the protein supplements had no effect on the diurnal variation in ruminal concentrations of peptide. Further carefully planned experiments are needed before the question can be answered.

MISCELLANEOUS EXPERIMENTS ON SYNCHRONIZATION OF ENERGY AND NITROGEN

Experiments have been reported (McCarthy, Klusmeyer, Vicini, Clark and Nelson, 1989; Herrera-Saldana, Gomez-Alarcon, Torabi and Huber,

1990a; Sinclair, Garnsworthy, Newbold and Buttery, 1993) that claim improvements in microbial protein synthesis attributable to improved synchrony of energy and nitrogen release in the rumen. However, as pointed out by others (Henning *et al.*, 1991; Robinson and McQueen, 1994), there is a serious problem of interpretation of these results, in that changes in synchronization have been achieved by manipulation of different dietary ingredients, thereby confounding synchronization with characteristics of the feeds. This point cannot be over emphasized. Whereas it is quite permissible to put forward effects on synchrony as a possible explanation of observed differences between dietary treatments, it is not permissible to use experiments of this type to test the hypothesis that synchrony affects microbial protein synthesis.

DIETS BASED ON GRASS SILAGE

Synchronization of energy and nitrogen release has been regarded as being particularly relevant to diets based on grass silage. It has long been the view that the release of ammonia from the substantial content of non-protein nitrogen (NPN) compounds in grass silage is very rapid, and this requires a similarly rapid release of energy in the rumen to ensure the most efficient microbial capture of ammonia. The greater output of microbial protein from the rumen when sugar, as opposed to starch, supplements are given (Table 1.1) appears to derive from the faster rate of fermentation of sugars. Based on the pattern of change of ruminal pH and the concentrations of total VFA, and with the exception of lactose, the sugars were fermented

Table 1.1   EFFECTS OF VARIOUS SUGARS AND MAIZE STARCH AS SUPPLEMENTS TO A BASAL DIET OF GRASS SILAGE ON RUMINAL DIGESTION AND THE OUTPUT OF MICROBIAL PROTEIN

		Supplement				
	Silage only	*Sucrose*	*Lactose*	*Xylose*	*Starch*	*Fructose*
Ruminal pH						
Mean	6.43	6.34	6.40	6.16	6.25	6.31
Minimum	6.04	5.72	6.05	5.97	5.99	5.74
NNH$_3$-N (mg/l)						
Mean	255	157	158	180	213	164
Maximum	354	240	234	240	288	233
Microbial CP*						
(g/d)	64	93	89	82	74	86

*Calculated from the urinary output of purine derivatives.
Chamberlain *et al.* (1993)

Table 1.2   EFFECTS OF MAIZE STARCH, COOKED MAIZE STARCH AND SUGAR
SUPPLEMENTS TO A BASAL DIET OF GRASS SILAGE ON RUMINAL DIGESTION AND
THE OUTPUT OF MICROBIAL PROTEIN

		Supplement			
	Silage only	*Starch*	*Cooked starch*	*Sucrose*	*Lactose*
Ruminal pH					
Mean	6.55	6.44	6.35	6.49	6.50
Minimum	6.43	6.39	6.07	6.24	6.24
NH_3-N (mg/l)					
Mean	268	233	205	217	231
Maximum	427	358	302	316	369
Protozoa ($\times 10^5$/ml)	2.7	7.1	6.4	4.1	4.0
Microbial CP* (g/d)	72	79	98	86	101

*Calculated from the urinary output of purine derivatives
Chamberlain, Choung and Robertson (unpublished)

more rapidly than the raw maize starch used in this experiment. Recent studies lend more support to this interpretation in that, when the starch was cooked before feeding, its effectiveness as a substrate for the production of microbial protein was increased to that of sugar (Table 1.2). Cooking maize starch had a marked effect on its rate of fermentation, as judged from its effects on ruminal pH; indeed, within the conditions of this experiment, its rate of fermentation was indistinguishable from that of the two sugars. These effects of cooking the starch are interesting also in that they lend no support to the earlier suggestion (Chamberlain, Thomas, Wilson, Newbold and MacDonald, 1985) that the inferior response to starch as opposed to sugar supplements was related to the increased numbers of protozoa that occur with starch feeding and the consequent increase in the intraruminal recycling of ammonia.

A closer examination of the increases in the output of microbial protein from the rumen in response to the addition of various carbohydrates (Table 1.3) reveals two main features. First, there are differences amongst carbohydrates, not only between starch and sugars, but also among the various sugars. Differences among the sugars may be related to the adverse effects on microbial growth of depressions of ruminal pH (Strobel and Russell, 1986) resulting from too rapid a rate of fermentation; there were suggestions of such effects with fructose. However, the lower response to xylose, evident in two studies, is not explicable in terms of a faster rate of fermentation, it being more slowly fermented than fructose or sucrose (Sutton,

1968), and must relate to an intrinsic difference in the microbial utilization of this sugar. The second feature that emerges from the table is that the responses to the carbohydrates are lower than the accepted maximum, especially in the studies with sheep. The suggestion is that the generally accepted low rates of microbial protein synthesis that occur with diets of silage only (Agricultural Research Council, 1984) are still evident in the incremental responses to the addition of fermentable carbohydrate. It is tempting to speculate that the rate of microbial growth is still constrained below maximum by deficiencies of other nutrients, in particular, amino acids and peptides. The very substantial response in the output of microbial protein to the continuous intraruminal infusion of casein (Rooke, Lee and Armstrong, 1987) would support this view but, on the other hand, responses to the addition of proteins to the diet have been inconsistent (Rooke, Bret, Overend and Armstrong, 1985; Dawson, Bruce, Buttery, Gill and Beever, 1988; Rooke and Armstrong, 1989; Beever, Gill, Dawson and Buttery, 1990). Obviously, further work is needed here but the inconsistent response to dietary protein supplements may reflect differences in the pattern of release of amino acids and peptides from different levels and types of protein.

Before jumping to the general conclusion that the synchronization of

Table 1.3 SYNTHESIS OF RUMINAL MICROBIAL PROTEIN (G/KG ADDED CARBOHYDRATE) FROM CARBOHYDRATE SUPPLEMENTS TO GRASS SILAGE

Carbohydrate	g microbial N/kg added carbohydrate	Reference
Sucrose	33	Huhtanen (1987a) (cattle)
Xylose	23	
Glucose syrup	28	Rooke *et al.* (1987) (cattle)
Glucose syrup	7	
plus casein	43	
Sucrose	21	Khaili and Huhtanen (1991 (cattle)
Sucrose	23	Chamberlain, Robertson and Choung (1993) (sheep)
Lactose	22	
Xylose	15	Chamberlain, Choung and Robertson (unpublished) (sheep)
Fructose	18	
Maize starch	9	
Cooked maize starch	21	
Sucrose	25	Chamberlain, Choung and Robertson (unpublished) (cattle)
Barley starch	20	
Cooked barley starch	28	

energy and nitrogen release can greatly affect the synthesis of microbial protein on diets containing high proportions of grass silage, it is well to consider the evidence, which boils down to: raw starch is a poorer substrate for the synthesis of microbial protein than is sugar or cooked starch. Note also that lactose and raw starch appear to have similar rates of fermentation (see later) and so a faster rate of fermentation of lactose cannot be the reason for the superiority of lactose over raw starch. It is clear that other factors are involved. Furthermore, to the extent that synchronization may be a factor to be considered, again there is no evidence that the matching of carbo-hydrate fermentation and ammonia release needs to be especially precise. Indeed, giving a supplement of sucrose either continuously or twice-daily resulted in a greater increase in output of microbial protein for the continuous treatment, even though a much closer matching of carbohydrate fermentation and ammonia release was obtained with the twice-daily treatment (Khalili and Huhtanen, 1989). As discussed by these workers, the twice-daily treatment may have resulted in less ATP yield per mole of sucrose fermented, since the acrylate pathway of propionate production can predominate at high rates of fermentation. There are inherent draw-backs in pursuing a very close matching of energy and nitrogen release because, even if it were necessary (and there is no evidence that it is), it can lead not only to a lower yield of ATP per mole of fermented carbohydrate with substrates such as sugars or cooked starches, but also to severe reductions of ruminal pH which have repercussions not only for microbial growth but also for the intake and digestibility of the diet.

Effects of RFC on fibre digestion and on the intake of forage

The depressions of fibre digestion, and related reductions of forage intake, that usually accompany the addition of significant amounts of RFC to the diet can be considered as made up of two components: a 'pH effect' and a 'carbohydrate effect' (Mould *et al.*, 1983). Thus we might assume that, for a given source of RFC, its 'pH effect' will be determined by its rate of fermentation and its 'carbohydrate effect' will depend on which microbial species utilize it and the interactions of these microbes with cellulolytic bacteria. It should be evident already that the 'carbohydrate effect' is not only difficult to define but is also likely to be even more difficult to measure. Again, even though the 'pH effect' might appear to lend itself more to measurement, it is important to remember that the depressions of ruminal pH resulting from the dietary inclusion of a given source of RFC will depend not only on its rate of fermentation but also on the rate of fermentation of the other components of the diet and the buffering capacity of the rumen. The picture becomes even more complex when we consider also that the 'pH effect' itself consists of at least two identifiable phases (Mould *et al.*, 1983; Hoover, 1986): one due to moderate depressions of pH, say between 6.8 and 6.2, which may relate to effects on microbial attachment to fibre; and one due to more severe depressions of pH which probably relates to severe inhibition of the growth of cellulolytic bacteria and reductions in their numbers.

It is established that replacing fibrous carbohydrate in the concentrate by starch reduces forage intake (see de Visser, 1993), but the question that concerns us here is whether different sources of RFC show different degrees of depression of forage intake. There is some evidence from experiments *in vitro* that the 'carbohydrate effect' of sugars is greater than that of starch (Simpson, 1984), which may derive from more species of cellulolytic bacteria being able to utilize soluble sugars than are able to utilize starch. Published reports of experiments *in vivo* involving direct comparisons of starch and sugars are rare, but the limited evidence would support the view that sugars have a greater intake-depressing effect than starch. Replacement of 33% of barley in an all-barley concentrate by molasses reduced silage intake by around 10% (Huhtanen, 1987b) and a concentrate containing 38% of sucrose reduced silage intake by around 12% compared with one based on barley (Chamberlain, Martin and Robertson, 1990). However, it should be emphasized that these comparisons apply only to barley starch and sucrose, and should not be regarded as a general statement applicable to all starches and all sugars: comparisons of cooked starches and more slowly fermented sugars (e.g. lactose) might well yield a different conclusion.

There is a further, important point to make in relation to the intake-depressing effects of RFC. The results of comparisons between different sources of RFC will depend very much on the dietary conditions in which the comparisons are made. This is well illustrated in the results of Chamberlain *et al.* (1990) referred to above (Table 1.4). Note that the reduction of silage intake induced by the high-sugar supplement was removed by the addition of a small amount of fish meal to the diet. The mechanism of action of the fish meal here is not known but it might well be a simple buffering effect on ruminal pH (Choung and Chamberlain, 1993a). The message from these observations is that the intake-depressing effects of RFC should be regarded more as potential effects rather than actual: there may well be simple remedies, such as changes in compound formulation to increase

Table 1.4 SILAGE INTAKE (kg DM/d) OF DAIRY COWS GIVEN 5kg/d OF A BARLEY-BASED OR HIGH-SUGAR CONCENTRATE WITH OR WITHOUT FISH MEAL

	Starch				*Sugar*			
Fishmeal (kg/d)	0	0.5	1.0	1.5	0	0.5	1.0	1.5
Intake	8.6	8.1	8.9	8.2	7.5	8.5	8.4	8.4

Chamberlain *et al.* (1990)

buffering capacity, that could substantially lessen adverse effects of the inclusion of more rapidly fermented forms of RFC.

Measurement of rate of fermentation

MEASUREMENT *IN VIVO*

The use of animals cannulated in the intestine can provide estimates of the extent of digestion of RFC in the rumen and in the intestines (see Theurer, 1986). However, these data are of limited use in distinguishing between different sources of RFC in terms of their rate of ruminal fermentation. Again, estimates based on the rate and extent of depression of ruminal pH or the rate of increase of VFA concentrations may provide a rough index of the rate of fermentation but, for reasons discussed earlier, such measurements are of limited accuracy, difficult to standardize and, in any case, do not lend themselves to routine use. At first sight, measurement of the rate of

Table 1.5 SOME PUBLISHED ESTIMATES OF SOLUBILITY IN BUFFER SOLUTIONS COMPARED WITH ZERO-TIME WASHOUT VALUES FROM NYLON-BAG INCUBATIONS OF STARCH FROM SOME CEREAL GRAINS

	Solubility (%)		*Zero-time loss (%)*	
	Buffer A	*Buffer B*	*1*	*2*
Maize	4	8	28	21
Milo	2	3	33	4
Wheat	—	2	69	78
Barley	5	11	65	66
Oats	5	4	96	97

Buffers: A, acetate (Herrera-Saldana *et al.* 1990b)
     B, bicarbonate-phosphate (Herrera-Saldana, Huber and Swingle, 1986)

Nylon-bag incubations: 1, machine-washing (Tamminga *et al.* 1990a)
     2, hand-washing (Herrera-Saldena *et al.* 1990b)

loss of RFC from feedstuffs held in nylon bags during incubation in the rumen (Herrera-Saldana, Huber and Poos, 1990b; Tamminga, Van Vuuren and Van Der Koelen, 1990) is an attractive option. However, there must be serious concern over the interpretation of the data, especially with respect to the fraction that is deemed to be instantly lost from the bag (zero time loss) and hence is classed as rapidly fermented. Examination of the data presented in Table 1.5 shows that the zero-time losses certainly do not represent material that is truly soluble, and must therefore be due to the loss of particulate matter. When this undefined fraction represents at least 60% of the starch in common cereals, it is difficult to see how this technique can provide any meaningful information on the rate of fermentation of different cereals, not to mention the effects of processing of the cereals. On a practical point, it should also be noted from Table 1.5 that the zero-time losses are not related to the method of washing the bags i.e. machine-washing versus hand-washing.

MEASUREMENT *IN VITRO*

Methods used in vitro break down essentially into two categories: those based on incubation with rumen fluid and those based on incubation with purified enzymes.

With methods based on incubation in rumen fluid, the first question is what to measure. Interest has focused mainly on the rates of production of VFA, gas or the rate of disappearance of substrate. Measuring the rate of gas production is relatively quick and easy. However, interpretation is not easy because the total production of gas derives from two sources: direct production of CO_2 and methane and indirect production of CO_2 from the action of VFA on bicarbonate in the rumen fluid and in the added buffer, and so varies with the mixture of VFA produced (Table 1.6). Hence, there is no clear relationship between the rate of fermentation and total gas production (Beuvink, 1993). On the other hand, even though the recovery of fermented sugars as VFA (and, where appropriate, lactic acid) is only around 40% (Sutton, 1968; Newbold and Rust, 1992), the rate of disappearance of sugar and the rate of appearance of fermentation acids are very highly correlated (Sutton 1968). However, most attention has been given to the fermentation of starch sources, and measurement of its disappearance from the incubation medium is easier than measuring VFA production. The results of extensive studies on the ruminal degradation of starch sources (Cone, ClineThiel, Marlestein and Van't Klooster., 1989; Cone, 1991) have shown that, although the actual values for disappearance rates

Table 1.6   DIRECT AND INDIRECT GAS PRODUCTION FROM 1 MOL OF
GLUCOSE FERMENTED TO DIFFERENT END PRODUCTS

Acidic end products (mol)	Direct gas (mol)	Indirect gas (mol)	Total gas (mol)
2 Acetic acid	$2\ CO_2$	$2\ CO_2$	$4\ CO_2$
1 Butyric acid	$2\ CO_2$	$1\ CO_2$	$3\ CO_2$
2 Propionic acid	—	$2\ CO_2$	$2\ CO_2$
2 Lactic acid	—	$2\ CO_2$	$2\ CO_2$

From Beuvink (1993)

vary with the diet of the donor animal, the ranking order of the starch
sources is little affected (Table 1.7). Furthermore, this method has been
shown to provide results that allow the effects of processing on the rate of
fermentation of starch to be clearly identified (Table 1.8).

The use of methods based on incubation with rumen fluid can pose
problems for some feed compounders. Consequently, attention has been
given to examining the rate of breakdown of starch using enzymes from
animal and microbial sources (Cone 1991). Although, of the purified
enzymes used, bacterial amylase was the most promising, its use has clear
limitations when it comes to its sensitivity to some of the effects of cereal
processing (Table 1.9). On the other hand, the results obtained using
reconstituted freeze-dried preparations of cell-free rumen fluid agree
reasonably well, in terms of the ranking order of the various maize products,
with those obtained from incubations with rumen fluid (Table 1.9). Further

Table 1.7   PERCENTAGE OF STARCH DEGRADED AFTER 6h INCUBATION
WITH RUMEN FLUID FROM A COW FED HAY (HFC) OR HAY PLUS TWO DIFFERENT
CONCENTRATES (CFC)

	HFC	CFC (A)	CFC (B)
Maize	3.5	21.6	21.6
Milocorn	5.5	23.6	25.1
Potato	6.6	25.1	36.5
Millet	6.8	25.7	31.0
Wheat	9.9	42.3	41.2
Barley	10.3	33.8	41.0
Oats	15.9	51.0	55.6
Tapioca	23.3	47.0	52.8
Paselli	55.9	60.4	58.9

Concentrates consisted mainly of tapioca and steam-flaked maize (A) and maize and milocorn (B).
From Cone *et al.* (1989)

Table 1.8   THE EFFECTS OF PROCESSING ON THE PERCENTAGE OF STARCH
DEGRADED AFTER INCUBATION FOR 6h WITH RUMEN FLUID FROM A COW FED
HAY (HFC) OR HAY PLUS TWO DIFFERENT CONCENTRATES

	HFC	CFC (A)	CFC (B)
Maize	3.5	21.6	21.6
Maize feed meal	11.1	36.1	36.1
Maize gluten feed	17.6	–	45.3
Steam flaked maize	14.4	34.9	42.6
Maize flake	15.8	38.5	47.7
Popped maize	31.7	49.8	52.2
Wheat	9.9	42.3	41.2
Wheat middlings	14.7	41.7	50.1
Wheat feed meal	31.3	43.5	54.9
Poppled wheat	44.0	57.8	66.4
Rice	7.8	22.4	25.3
Rice feed meal	21.8	45.3	43.4
Popped rice	50.3	57.6	56.9

Concentrates consisted mainly of tapioca and steam-flaked maize (A) and maize and milocorn (B).
From Cone *et al.* (1989)

development of this approach could provide a useful method for routine use
by feed compounders.

When it comes to assigning values for the rate of fermentation of sources
of RFC for rationing purposes, the best that current knowledge will allow is
the construction of a provisional 'Fermentation Rate Index' derived from
measurements *in vitro*. However, even this is difficult because there is
insufficient information on the comparative rates of fermentation of the
various starch and sugar sources. Depression of pH during incubation with
rumen liquor *in vitro* was closely correlated with the production of total
fermentation acids (Sutton, 1968). Under suitably controlled conditions,
the rate of depression of pH might provide a useful index of the rate of
fermentation. Some results from a preliminary investigation along these
lines are shown in Table 1.10. The time course of changes in pH was
followed over 3.5 h of incubation, during which time the depression of pH
was virtually linear. Results are expressed relative to the rate of depression
of pH observed for glucose as a substrate. Although the method appears
promising, the provisional nature of these results must be emphasized;
further more detailed experiments are needed to confirm the validity of this
approach. However, it is interesting to note that the similarity in the
predicted rates of fermentation of lactose and raw starch agrees with rates of
depression of ruminal pH seen *in vivo* for these two substrates (Chamberlain

Table 1.9   THE PERCENTAGE OF STARCH DEGRADED AFTER 6h INCUBATION
IN RUMEN FLUID OR IN A CELL-FREE PREPARATION OF THE RUMEN FLUID
(FREEZE-DRIED AND RECONSTITUTED) OR AFTER 4h INCUBATION WITH
BACTERIAL AMYLASE

	Rumen fluid	Bacterial amylase	Cell-free preparation
Maize	19.6	23.2	8.4
Maize feed meal	44.0	34.8	35.6
Maize gluten feed	37.9	52.3	41.8
Steam flaked maize	34.1	77.1	44.3
Maize flake	51.9	77.1	65.1
Popped maize	74.9	74.6	82.0

From Cone (1991)

et al., 1993). This makes it difficult to argue for the superiority of lactose over
raw starch, as a supplement to grass silage (Tables 1.1 and 1.2) being due to
a faster rate of fermentation of lactose.

RATE OF FERMENTATION AND THE MOLAR COMPOSITION OF
VFA PRODUCED

Any assessment of the rate of fermentation of energy sources should go
hand in hand with a consideration of the molar composition of the mixture

Table 1.10   THE RATE OF DEPRESSION OF pH DURING INCUBATIONS *IN
VITRO* WITH RUMEN LIQUOR FOR SOME PURIFIED SUBSTRATES AND SOME
FEEDS. VALUES ARE EXPRESSED RELATIVE TO THAT OF GLUCOSE (100)

Purified substrates		Feeds	
Sucrose	100	Maize meal	49
Lactose	58	Cooked maize	78
Xylose	90	Wheat	56
Xylan	42	Cooked Wheat	88
Pectin	94	Barley	54
Maize starch	52		
Cooked maize starch	80		
Barley starch	56		
Cooked barley starch	90		

Chamberlain and Choung (unpublished)

of VFAs produced from their fermentation. This is obviously important because of effects of the molar ratio of the absorbed mixture of VFA on milk composition (see Thomas and Chamberlain, 1984), but it may be import-ant also because of interactions between the metabolism of propionate and ammonia in the liver, that have come to light more recently. Particularly when there is a high rate of absorption of ammonia from the rumen in the early post-feeding period, propionate can reduce the capacity of the liver to detoxify the ammonia via the urea cycle, with the result that ammonia spills over into peripheral blood, leading to effects on insulin secretion, with implications for the partition of nutrients between the udder and other body tissues, and possibly also for milk composition (Choung and Chamberlain, 1995). These findings would be expected to be of relevance to the feeding of diets containing high proportions of grass silage, especially when the silage has a high concentration of crude protein. Note that these are precisely the conditions in which attempts to synchronize energy and nitrogen release in the rumen are most applicable.

Whilst there is no doubt that specific sugars are fermented to mixtures of VFA that are broadly characteristic of them as substrates (Sutton, 1968; Chamberlain *et al.*, 1993), there remains uncertainty over the extent to which such factors as the nature of the basal diet and the level of addition of the sugars to the rumen can influence the composition of the VFA mixture produced. It has been proposed (Prins, Lankhorse and Van Hoven, 1984) that the acetate: propionate ratio is controlled by the hexose flux through the cell, and that the pattern of VFA production from a given source and amount of RFC would depend on the concentration of bacteria in the rumen; the lower the number of bacteria, the higher the flux per cell. If this is true, it offers a plausible explanation of effects of the basal diet, the level of addition of RFC and the frequency of feeding of diets rich in RFC (see Sutton, 1981). Again, the hexose flux through the microbial cell would be strongly affected by the rate at which starch is degraded in the rumen, which, in turn, is dependent on the source of the starch and its treatment during processing. Hence, it is not possible, with present knowledge, to generalize about the effects of the dietary inclusion of starches or sugars on milk composition. Whilst it is known, for example, that the severity of depressions of milk fat concentration varies with the inclusion of different types of starch (see Sutton, 1981), there are no reports of systematic and sufficiently comprehensive studies of the effects of source of starch and processing treatments on milk composition; the beneficial effects that starch inclusion can have on the concentration of milk protein (Sutton, Morant, Bines, Napper and Givens, 1993) serves to emphasize the need for such studies.

Conclusions

There is no convincing evidence of a need for close synchronization of energy and nitrogen release in the rumen to ensure efficient synthesis of microbial protein. This is not to say that severe mismatching of energy and nitrogen release may not have detrimental effects on microbial protein synthesis in some dietary circumstances, notably where a rapid release of ammonia immediately after feeding is coupled with the only significant source of energy being derived from slowly fermented (usually fibrous) substrates. Such a combination would be expected to result in ruminal pH being maintained at relatively high levels, which would favour the absorption of substantial amounts of ammonia from the rumen. In such conditions, the dietary inclusion of an appropriate source of RFC would be expected to improve the efficiency of microbial capture of ammonia. However, the different effects that various types of RFC can have on microbial protein synthesis in such conditions are not explicable in terms of improved synchronization of energy and ammonia availabilities; experimental results suggest the involvement of effects of specific substrates on the activities of the microbial population, but the nature of these effects remains to be established.

Whilst it is prudent, when formulating rations, to avoid very marked mismatches in the rates of ruminal release of energy and nitrogen, there are clear potential dangers, associated with inducing wide fluctuations of ruminal pH and even ruminal acidosis, in giving too much weight to achieving close synchronization if this entails incorporating substantial amounts of certain sources of RFC. Indeed, the best approach might be to attempt to achieve the most even pattern of energy supply in the rumen (Henning, Steyn and Meissner, 1993).

To what extent the availability of amino acids and/or peptides in the rumen can limit microbial protein synthesis remains an important question that can only be answered after further experimentation. At this stage, it is worth noting that culture studies *in vitro* indicate a clear, specific requirement of amylolytic bacteria for preformed amino acid/peptide. Although the full implications await experimentation *in vivo*, we would do well to recognize that the nature of the energy source in the diet will influence the nutritional requirements of the microbial population in the rumen, and that the level and type of dietary protein may need to be adjusted (on the basis of considerations that go beyond the scope of the current MP system), to ensure that the requirements for maximal rates of microbial protein synthesis are met. We should remember, also, that limitations in the ruminal

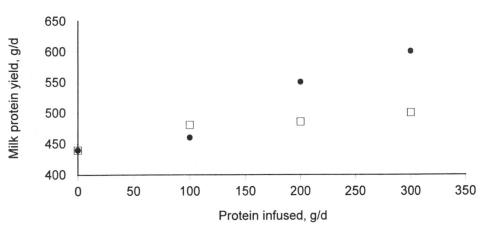

Figure 1.3 The responses in the yield of milk crude protein to the infusion of soya-protein isolate (□) or casein (●) into the abomasum in dairy cows receiving a grass silage and a cereal-based concentrate. (Choung and Chamberlain, 1993b)

supply of amino acid/peptide are likely to be particularly relevant to diets containing high proportions of grass silage.

Related to the whole issue under discussion in this paper is the question: even if modest increases in the yield of microbial protein can be achieved by manipulation of compound formulation, what would be the response of milk production? Although there is no clear answer to this question at present, it is worth considering the results of recent experiments in which proteins have been infused into the abomasum in dairy cows. Note the linear increase in the yield of milk protein in response to the infusion of casein, but the very limited increase in protein yield in response to the infusion of soya protein (Figure 1.3). When we consider that microbial protein would have a lower digestibility than soya-protein isolate and, probably, a rather similar biological value, the implication of results such as these is, at the very least, to seriously question any assumption that benefits to milk production would follow, as a matter of course, from an increased ruminal output of microbial protein.

Leaving aside the issue of synchronization, information on the rate and pattern of fermentation of carbohydrate sources is urgently needed if we are to uncover the mechanisms underlying the effects of diet on metabolism, and hence on milk composition, and use them to advantage in feeding the dairy cow.

Acknowledgements

We are grateful to Dr A.G. Williams for stimulating and helpful discussions.

References

Agricultural Research Council (1984) *The Nutrient Requirements of Ruminant Livestock* Suppl no 1 Slough: Commonwealth Agricultural Bureaux

Beever, D.E., Gill, M., Dawson, J.M. and Buttery, P.J. (1990) The effect of fish meal on the digestion of grass silage by growing cattle. *British Journal of Nutrition,* **63**, 489–502

Beever, D.E. and Cottrill, B.R. (1994) Protein systems for feeding ruminant livestock: a European assessment. *Journal of Dairy Science,* **77**, 2031–2043

Beuvink, J.M.W. (1993) Measuring and modelling in vitro gas production kinetics to evaluate ruminal fermentation of feedstuffs. Thesis, Lelystad, Netherlands: Institute for Animal Nutrition

Broderick, G.A., Wallace, R.J. and Ørskov, E.R. (1991) Control of rate and extent of protein degradation. In *Physiological Aspects of Digestion and Metabolism in Ruminants,* pp. 541–592 Edited by T. Tsuda, Y. Sasaki and R. Kawashima. London: Academic Press

Bryant, M.P. (1973) Nutritional requirements of the predominant rumen cellulolytic bacteria. *Federation Proceedings,* **32**, 1809–1813

Chamberlain, D.G., Thomas, P.C., Wilson, W.D., Newbold, C.J. and MacDonald, J.C. (1985) The effects of carbohydrate supplements on ruminal concentrations of ammonia in animals given diets of grass silage. *Journal of Agricultural Science, Cambridge,* **104**, 331–340

Chamberlain, D.G., Martin, P.A. and Robertson, S. (1990) The influence of the type of carbohydrate in the supplement on responses to protein supplementation in dairy cows receiving diets containing a high proportion of grass silage. Research Meeting No 2, British Grassland Society, Section VII, Poster 2, The British Grassland Society: Maidenhead

Chamberlain, D.G., Robertson, S. and Choung, J.-J. (1993) Sugars versus starch as supplements to grass silage: Effects on ruminal fermentation and the supply of microbial protein to the small intestine, estimated from the urinary excretion of purine derivatives, in sheep. *Journal of the Science of Food and Agriculture,* **63**, 189–194

Choung, J.-J. and Chamberlain, D.G. (1993a) Effects of addition of lactic acid and post-ruminal supplementation with casein on the nutritional value of grass silage for milk production in dairy cows. *Grass and Forage Science,* **48**, 380–386

Choung, J.-J. and Chamberlain, D.G. (1993b) The effects of abomasal infusions of casein or soya-bean–protein isolate on the milk production of dairy cows in mid-lactation. *British Journal of Nutrition,* **69**, 103–115

Choung, J.-J. and Chamberlain, D.G. (1995) Effects of ruminal infusion of propionate on the concentrations of ammonia and insulin in peripheral blood of cows receiving an intraruminal infusion of urea. *Journal of Dairy Research* (in press).

Cone, J.W., Cline-Theil,W., Malestein, A. and van't Klooster, A.Th. (1989) Degradation of starch by incubation with rumen fluid. A comparison of different starch sources. *Journal of the Science of Food and Agriculture,* **49**, 173–183

Cone, J.W. (1991) Degradation of starch in feed concentrates by enzymes, rumen fluid and rumen enzymes. *Journal of the Science of Food and Agriculture,* **54**, 23–34

Dawson, J.M., Bruce, C.I., Buttery, P.J., Gill, M. and Beever, D.E. (1988) Protein metabolism in the rumen of silage-fed steers: effects of fish meal supplementation. *British Journal of Nutrition,* **55**, 339–353

de Visser, H. (1993) Characterization of carbohydrates in concentrates for dairy cows. In *Recent Advances in Animal Nutrition – 1993,* pp. 19–38. Edited by P.C. Garnsworthy and D.J.A. Cole. Nottingham: Nottingham University Press

Henning, P.H., Steyn, D.G. and Meissner, H.H. (1991) The effect of energy and nitrogen supply pattern on rumen bacterial growth in vitro. *Animal Production,* **53**, 165–175

Henning, P.H., Steyn, D.G. and Meissner, H.H. (1993) Effect of synchronization of energy and nitrogen supply on ruminal characteristics and microbial growth. *Journal of Animal Science,* **71**, 2516–2528

Herrera-Saldana, R., Huber, J.T. and Swingle R.S. (1986) Protein and starch solubility and degradability of several common feedstuffs. *Journal of Dairy Science,* **69** (Suppl 1), 141 (Abstr)

Herrera-Saldana, R., Gomez-Alarcon, R., Torabi, M. and Huber, J.T. (1990a) Influence of synchronizing protein and starch degradation in the rumen on nutrient utilization and microbial protein synthesis. *Journal of Dairy Science,* **73**, 142–148

Herrera-Saldana, R., Huber, J.T. and Poos, M.H. (1990 b) Dry matter, crude protein and starch degradability of five cereal grains. *Journal of Dairy Science,* **73**, 2386–2393

Hespell, R.B. (1984) Influence of ammonia assimilation pathways and survival strategy on rumen microbial growth. In *Herbivore Nutrition,* pp. 346–358 Edited by F.M.C. Gilchrist and R.I. Mackie. Craighall, S. Africa: The Science Press

Hespell, R.B. and Bryant, M.P. (1979) Efficiency of rumen microbial growth: influence of some theoretical and experimental factors on YATP. *Journal of Animal Science*, **49**, 1641–1647

Hoover, W.H. (1986) Chemical factors involved in ruminal fibre digestion. *Journal of Dairy Science*, 69, 2755–2766

Hoover, W.H. and Stokes, S.R. (1991) Balancing carbohydrates and proteins for optimum rumen microbial yield. *Journal of Dairy Science*, **74**, 3630–3644

Huhtanen, P. (1987a) The effects of intraruminal infusions of sucrose and xylose on nitrogen and fibre digestion in the rumen and intestines of cattle receiving diets of grass silage and barley. *Journal of Agricultural Science in Finland*, **59**, 405–424

Huhtanen, P. (1987b) The effect of dietary inclusion of barley, unmolassed sugar beet pulp and molasses on milk production, digestibility and digesta passage in dairy cows given silage-based diet. *Journal of Agricultural Science in Finland*, **59**, 101–120

Khalili, H. and Huhtanen, P. (1991) Sucrose supplements in cattle given a grass silage-based diet. 1. Digestion of organic matter and nitrogen. *Animal Feed Science and Technology*, **33**, 247–261

Knight, W.M. and Owens, F.N. (1973) Interval urea infusion for lambs. *Journal of Animal Science*, **36**, 145–149

McCarthy, R.D., Klusmeyer, T.H., Vicini, J.L., Clark, J.H. and Nelson, D.R. (1989) Effects of source of protein and carbohydrate on ruminal fermentation and passage of nutrients to the small intestine of lactating cows. *Journal of Dairy Science*, **72**, 2002–2010

Mackie, R.I. and Therion, J.J. (1984) Influence of mineral interactions on growth efficiency of rumen bacteria. In *Herbivore Nutrition*, pp. 455–477 Edited by F.M.C. Gilchrist and R.I. Mackie. Craighall, S. Africa: The Science Press,

Meggison, P.A., Mcmeniman, N.P. and Armstrong, D.G. (1979) Efficiency of utilization of non-protein nitrogen in cattle. *Proceedings of the Nutrition Society*, **38**, 147A

Mould, F.L., Ørskov, E.R. and Mann, S.O. (1983) Associative effects of mixed feeds. 1. Effects of type and level of supplementation and the influence of the rumen fluid pH on cellulolysis in vivo and dry matter digestion of various roughages. *Animal Feed Science and Technology*, 10, 15–30

Newbold, J.R. (1994) Practical application of the Metabolizable Protein System. In *Recent Advances in Animal Nutrition 1994*, pp. 231–264. Edited by P.C. Garnsworthy and D.J.A. Cole. Nottingham: Nottingham University Press

Newbold, J.R. and Rust, S.R. (1992) Effect of asynchronous nitrogen and energy supply on growth of ruminal bacteria in batch culture. *Journal of Animal Science*, **70**, 538–546

Obara, Y., Dellow, D.W. and Nolan, J.V. (1991) The influence of energy-rich supplements on nitrogen kinetics in ruminants. In *Physiological Aspects of Digestion and Metabolism in Ruminants*, pp. 515–539. Edited by T. Tsuda, Y. Sasaki and R. Kawashima. London: Academic Press

Prins, R.A., Lankhorse, A. and van Hoven, W. (1984) Gastrointestinal fermentation in herbivores and the extent of plant cell-wall digestion. In *Herbivore Nutrition in the Subtropics and Tropics*, pp. 408–434. Edited by F.M.C. Gilchrist and R.I. Mackie. Craighall, South Africa: The Science Press

Robinson, P.H. and McQueen, R.E. (1994) Influence of supplemental protein source and feeding frequency on rumen fermentation and performance in dairy cows. *Journal of Dairy Science*, **77**, 1340–1353

Rooke, J.A., Brett, P.A., Overend, M.A. and Armstrong, D.G. (1985) The energetic efficiency of rumen microbial protein synthesis in cattle given silage-based diets. *Animal Feed Science and Technology*, **13**, 255–267

Rooke, J.A., Lee, N.H. and Armstrong, D.G. (1987) The effects of intraruminal infusions of urea, casein, glucose syrup and a mixture of casein and glucose syrup on nitrogen digestion in the rumen of cattle receiving grass silage diets. *British Journal of Nutrition*, **57**, 89–98

Rooke, J.A. and Armstrong, D.G. (1989) The importance of the form of nitrogen on microbial protein synthesis in the rumen of cattle receiving grass silage and continuous intraruminal infusions of sucrose. *British Journal of Nutrition*, **61**, 113–121

Russell, J.B., Sniffen, C.J. and van Soest, P.J. (1983) Effect of carbohydrate limitation on degradation and utilization of casein by mixed rumen bacteria. *Journal of Dairy Science*, **66**, 763–770

Russell, J.B and Wallace, R.J. (1988) Energy yielding and consuming reactions. In *The Rumen Microbial Ecosystem*, pp. 185–216 Edited by P.N. Hobson. London: Elsevier

Russell, J.B, O'Connor, J.D., Fox, D.G., van Soest, P.J. and Sniffen, C.J. (1992) A net carbohydrate and protein system for evaluating cattle diets. 1. Ruminal fermentation. *Journal of Animal Science*, **70**, 3551–3561

Salter, D.N., Smith, R.H. and Hewitt, D. (1983) Factors affecting the capture of dietary nitrogen by microorganisms in the forestomachs of the young steer. Experiments with [15-N]urea. *British Journal of Nutrition*, **50**, 427–435

Simpson, M.E. (1984) Protective effect of alternative carbon sources towards cellulose during its digestion in bovine rumen fluid. *Developments in Industrial Microbiology*, **25**, 641–649

Sinclair, L.A., Garnsworthy, P.C., Newbold, J.R. and Buttery, P.J. (1993) Effect of synchronizing the rate of dietary energy and nitrogen release on rumen fermentation and microbial protein synthesis in sheep. *Journal of Agricultural Science, Cambridge*, **120**, 251–263

Smith, R.H. (1975) Nitrogen metabolism in the rumen and the composition and nutritive value of nitrogen compounds entering the duodenum. In *Digestion and Metabolism in the Ruminant*, pp. 399–415. Edited by I.W. McDonald and A.C.I. Warner. Armidale: University of New England Publishing Unit

Stewart, C.S., Paniagua, C., Dinsdale, D., Cheng, K-J. and Garrow, S.H. (1981) Selective isolation and characteristics of Bacteroides succinogenes from the rumen of a cow. *Applied and Environmental Microbiology*, **41**, 504–510

Stouthamer, A.H. (1973) A theoretical study on the amount of ATP required for synthesis of microbial cell material. *Antonie van Leeuwenhoek*, **39**, 545–565

Streeter, C.L., Little, C.O., Mitchell, G.E. and Scott, R.A. (1973) Influence of rate of ruminal administration of urea on nitrogen utilization in lambs. *Journal of Animal Science*, **37**, 796–799

Strobel, H.J. and Russell, J.B. (1986) Effect of pH and energy spilling on bacterial protein synthesis by carbohydrate-limited cultures of mixed rumen bacteria. *Journal of Dairy Science*, **69**, 2941–2947

Sutton, J.D. (1968) The fermentation of soluble carbohydrates in rumen contents of cows fed diets containing a large proportion of hay. *British Journal of Nutrition*, **22**, 689–712

Sutton, J.D. (1981) Concentrate feeding and milk composition. In *Recent Advances in Animal Nutrition – 1981*, pp. 35–48 Edited by W. Haresign. London: Butterworths

Sutton, J.D., Morant, S.V., Bines, J.A., Napper, D.J. and Givens, D.I. (1993) Effect of altering the starch:fibre ratio in the concentrates on hay intake and milk production by Friesian cows. *Journal of Agricultural Science, Cambridge*, **120**, 379–390

Tamminga, S., van Vuuren, A.M., van der Koelen, C.J. (1990) Ruminal behaviour of structural carbohydrates, non-structural carbohydrates and crude protein from concentrate ingredients in dairy cows. *Netherlands Journal of Agricultural Science*, 38, 513–526

Theurer, C.B. (1986) Grain processing effects on starch utilization by ruminants. *Journal of Animal Science*, **63**, 1649–1662

Thomas, P.C. and Chamberlain, D.G. (1984) Manipulation of milk composition to meet market needs. In *Recent Advances in Animal Nutrition –
1984*, pp. 219–243 Edited by W. Haresign and D.J.A. Cole. London:
Butterworths

2

THE USE OF BLOOD BIOCHEMISTRY FOR DETERMINING THE NUTRITIONAL STATUS OF DAIRY COWS

W.R. WARD, R.D. MURRAY, A.R. WHITE[a] and E.M. REES
University of Liverpool, Leahurst, Neston, South Wirral, L64 7TE, UK
[a] AF plc, Kinross, New Hall Lane, Preston PR1 5JX, UK

Introduction

METABOLIC PROFILES

Metabolic profiles in dairy cattle were made popular by the late Jack Payne in the 1970's at the Compton Laboratory (Payne, Dew, Manston and Faulks, 1970). Despite the authors' warning that 'It must be plainly stated that the metabolic profile test is merely an aid to preventive veterinary medicine', it disappointed those who expected it to revolutionise dairy cow nutrition. The choice of metabolites may have owed more to the repertoire of the Autoanalyzer than the usefulness to the veterinarian and cattle nutritionist. The normal ranges for the metabolites were based on the standard deviations around the mean found in the herds originally tested, and believed to be normal. The early results showed interesting variations in metabolites between seasons, and during the reproductive cycle of the cow, but some of these showed no obvious relationship to subclinical problems.

Since fertility is the biggest single source of loss to dairy farmers, it is not surprising that metabolic profiles were soon investigated for their ability to detect abnormalities in the diet related to fertility. Parker and Blowey (1976) at the Central Veterinary Laboratory found that in the herds that they examined the blood measurements showed no consistent relationship either to the estimated feed intake, or to the fertility.

Early studies (Lewis, 1957) showed that blood urea concentration was closely linked to the concentration of ammonia in the portal circulation and was related to the concentration of ammonia in the rumen 4 to 6 hours earlier.

Variation in blood composition at different times of day is clearly a factor limiting the accuracy of measurement of some metabolites. Manston, Rowlands, Little and Collis (1981) showed that out of 13 metabolites betahydroxybutyrate fluctuated most, varying 2.5 fold between 0700 and 1100 hrs in lactating cows only. Diurnal variation was also measured in urea, magnesium and copper, with coefficients of variation of 18%, 7% and 16% respectively. Gustafsson and Palmquist (1993) found that in individual cows blood urea concentration rose from 6 to 8 or 9 mmol/l following a single daily feed. Pearson, Craig and Rowe (1992) found that bile acids varied by up to 60 mmol/l within an hour.

THE DAIRY HERD MONITORING SCHEME

The University of Edinburgh and Dalgety began a long relationship, which is still in being. In 1991 the University of Liverpool, together with the then Amalgamated Farmers (now AF plc) set up, after a pilot study, the Dairy Herd Monitoring Scheme (DHMS). The aims of the scheme included 'To link, through the means of planned blood testing, nutrition and preventive medicine on selected dairy units. To allow a greater degree of dialogue between the AF representative and the local veterinary practice. To collate information so that it might be used by either the University of Liverpool or by Amalgamated Farmers...'.

From the beginning, the scheme was seen as a planned exercise. AF cattle specialists selected farmers who were willing to commit themselves to, and to pay for, a series of four sets of sampling over one year. Farmers had to select cattle at the correct stage of lactation or dry period, at the correct time relative to changes in the feed. A considerable amount of information about the herd and the individual cows tested, had to be submitted with the samples. A copy of the Input Form is shown as Figure 2.1.

One important aim of the exercise was to involve veterinarians in discussion of nutrition with farmers and nutritionists.

Blood was to be taken from 6 cows in early lactation, 5 in mid-lactation, and 6 dry cows. Three vacutainers were used, oxalate/fluoride for glucose, heparin for glutathione peroxidase and copper, and plain for other metabolites. The metabolites measured, the methods and the standard values are shown in Table 2.1. Metabolites were measured using a Specific Selective Chemistry Analyser with Supra software (Kone, Ruukintie, Finland) and kits (Randox Laboratories, Crumlin, Northern Ireland) in serum, or in whole blood for glutathione peroxidase: copper was measured in plasma, using atomic absorption spectrophotometry. Glutathione peroxidase was

DEPARTMENTS OF VETERINARY CLINICAL SCIENCE
AND VETERINARY PATHOLOGY

92ᴸ – 2292

DAIRY HERD MONITORING SCHEME
INPUT FORM

CODE No. O1 - 09 - 03 - 02   DATE 3 8 92

A **COW INFORMATION**								
Cow No.	Calving Date	Lact No.	Yield (kg)	Pred. yld (kg/lact)	App.Wt. (kg)	Cond. Sc.	Parl.Feed (kg/day)	Comments
E 719	3.7.92	2	30.0		580	2½	5.0	✱ ALL SAMPLES TAKEN
A 807	5.7.92	1	20.0		570	2½	5.0	FROM JUGULAR VEIN
R 625	15.7.92	2	35.2		600	3	5.0	
L 591	5.7.92	4	29.8		610	3½	5.0	
Y 723	1.7.92	3	27.0		580	2½	5.0	
L 650	6.7.92	3	35.2		585	3	5.0	
A 658	29.6.92	4	35.0		610	3	5.0	
C **T.** 739	7.7.92	1	25.0		600	2½	5.0	HOME MIX
M 814	14.7.92	1	20.0		575	2	5.0	
I **D** 690	13.7.92	3	32.0		575	3	5.0	SUGAR BEET 48%
								MAIZE GLUTEN 24%
L 728	9.8.92	2			570	3	-	SOYA 10%
A **C** **T.** 733	22.9.92	2			540	2	-	MOLASSES 12%
702	11.8.92	2			685	3½	-	FISH 3%
257	8.8.92	10			685	4½	-	MINS/VIT 1%
D 544	14.8.92	5			730	5	-	
R **Y** 489	6.8.92	5			700	4½	-	
775	20.9.92	2			560	2½	-	

B **FEED INFORMATION**	D.M.	ME	CP	DCP	pH	'D'	NH₃N	ESTIMATED INTAKES EARLY	MID	DRY
SILAGE	20	10.6	155	103	3.7		8.0	AD LIB		-
HOME MIX								5.0		-
GRASS								AD LIB		AD LIB
DAIRYLINE 16 THROUGH PARLOUR										

C **MANAGEMENT INFORM.**	(tick where appropriate) ✱ PLEASE ANALYSE FOR GSH·Px
	& COPPER

ALL COWS

Self Feed	Easy Feed	'Lead Feed
Forage Box	Flat Rate	Complete Diet
Group Feeding		

Figure 2.1 Dairy herd monitoring scheme input form.

Table 2.1   BLOOD METABOLITES MEASURED, METHODS, AND STANDARD VALUES

Metabolite	Method	Standard value
Energy		
Betahydroxybutyrate (BOHB)	Ranbut*	< 0.9 mmol/l
Glucose	GOD/PAP*	> 2.5 mmol/l
Protein		
Urea	Enzymatic kinetic*	3.3 to 5 mmol/l
Albumin	Bromocresol green*	30 to 40 g/l
Total protein	Biuret*	60 to 80 g/l
Globulin	Difference	30 to 40 g/l
Minerals		
Magnesium	Col'metric xylidine blue	> 0.74 mmol/l
Phosphate (inorganic)	UV*	> 1.8 mmol/l
(phosphate in jugular blood)		> 1.6 mmol/l
Glutathione peroxidase (selenium)	Ransel*	> 39 units/33% PCV
Copper	Atomic absorption	> 9.4 micromol/l
(copper in serum)		> 7.4 micromol/l
Liver		
Bile acids	Enzymatic colorimetric*	< 60 micromol/l

* Randox Laboratories Ltd, Diamond Road, Crumlin, Co Antrim, BT29 4QY, UK

measured as an indicator of selenium status (Anderson, Berrett and Patterson, 1978). The concentration of glutathione peroxidase was calculated assuming a packed cell volume of 33%. External quality control (QC) was conducted by Randox Laboratories.

Other metabolites were measured on request: the most common requests have been for cobalt, when cyanocobalamin has been measured, and iodine, when thyroxine (T4) has been used as the best available measure.

Table 2.2   GEOGRAPHICAL DISTRIBUTION OF FARMS

Lancashire	31	Northumberland	6
Scotland	22	Cheshire	5
Lincolnshire	21	Hereford & Worcs	4
Yorkshire	20	Nottinghamshire	4
Cumbria	15	Cambridgeshire	1
Derbyshire	8	Humberside	1
Shropshire	6	Leicestershire	1

Samples were almost always analysed on the day of receipt, and copies of results along with a commentary compiled by a university veterinarian (WRW or RDM) were sent to the farmer, veterinary surgeon and AF nutritionist.

Results

OVERVIEW

A total of 145 farms registered for the scheme, in addition to those that sent single batches of samples. Most farms were in the North and Midlands of England, with 22 in Scotland (Table 2.2).

A total of 10,199 cows were tested up to September 1994. Over 1,000 cows were sampled in each lactation up to lactation 5. Cows in first lactation (heifers) accounted for 13.5% of all the cows sampled (Table 2.3). Lactation number was not recorded in 160 cases.

Table 2.3   DISTRIBUTION OF COWS BY LACTATION

Lactation	Number	Percentage
1	1358	13.5%
2	2072	20.6%
3	1844	18.4%
4	1492	14.9%
5	1152	11.5%
6	863	8.6%
> 6	1258	12.5%

Table 2.4   NUMBER OF COWS AT DIFFERENT STAGES OF LACTATION

Stage	Days post-partum	Number
1. Postpartum	1 to 41	2198
2. Early	42 to 120	2217
3. Mid	121 to 199	1807
4. Late	200 to 325	1517
5. Dry	305 to calving	2460

Table 2.5   NUMBER OF COWS CALVING IN EACH MONTH

Jan	Feb	Mar	Apr	May	Jun	July	Aug	Sep	Oct	Nov	Dec
867	667	723	549	466	690	929	1067	1076	1088	945	1047

In 85 cases the month of calving was not given.

When lactation was divided into four stages plus the dry period, the largest group comprised the dry cows, with 2,500 cows. The smallest category consisted of the cows in late lactation (Table 2.4).

The number of cows calving in each month shows a peak in August to October, and a trough in May (Table 2.5).

About half of all dry cows were scored above condition score 3, and only 3% of dry cows were below condition score 2. In the group of cows calved 6 to 17 weeks, (Early group) 30% were below condition score 2.5 (Table 2.6).

The daily milk yield at the time of sampling was 30 to 39 kg in over 40% of cows calved up to 6 weeks (Post-partum group) and in cows calved 6 to 17 weeks (Early group), and 20 to 29 kg in 35 to 40% of these two groups (Table 2.7).

The amount of parlour compound being fed at the time of sampling ranged from under 1 to over 12 kg (Table 2.8). The mode was 6 to 6.9 kg in cows calved up to 6 weeks (Post-partum) and in cows calved 6 to 17 weeks

Table 2.6   NUMBER AND PERCENTAGE OF COWS IN EACH BAND OF CONDITION

Category	< 2.0		2 to 2.4		2.5 to 3		> 3		Total
Postpartum	108	(4.9%)	401	(18.2%)	1233	(56%)	456	(20.7%)	2198
Early	195	(8.8%)	520	(23.5%)	1165	(25%)	338	(15.2%)	2217
Mid	102	(5.6%)	329	(18.2%)	1039	(58%)	337	(18.6%)	1807
Late	74	(4.9%)	201	(13.2%)	743	(49.0%)	499	(32.9%)	1517
Dry	54	(2.2%)	76	(3.1%)	941	(38.3%)	1389	(57.0%)	2460

Table 2.7   NUMBER AND PERCENTAGE OF COWS WITH DAILY MILK YIELD AT TIME OF SAMPLING IN EACH CATEGORY

	< 19 kg		20 to 29 kg		30 to 39 kg		> 40 kg		Total
P	345	(15.8%)	782	(35.9%)	901	(41.3%)	151	(6.9%)	2179
E	217	(9.9%)	861	(39.3%)	957	(43.7%)	155	(7.1%)	2190
M	495	(28.6%)	958	(55.4%)	259	(15.0%)	17	(1.0%)	1729
L	992	(68.4%)	413	(28.5%)	44	(3.0%)	0		1449

Table 2.8 NUMBER AND PERCENTAGE OF COWS RECEIVING PARLOUR COMPOUND OF DIFFERENT AMOUNTS AT TIME OF SAMPLING

Parlour cpd Kg/day	Post partum		Early		Mid		Late		Dry	
< 1	10	(0.6%)	10	(0.5%)	35	(2.2%)	64	(6.7%)	44	(90.6%)
1 to 1.9	29	(1.6%)	82	(4.2%)	169	(10.7%)	248	(26.1%)	111	(4.5%)
2 to 2.9	100	(5.5%)	134	(1.8%)	258	(16.3%)	209	(22.0%)	49	(2.0%)
3 to 3.9	136	(7.6%)	182	(9.4%)	222	(14%)	111	(11.7%)	19	(0.7%)
4 to 4.9	215	(12.0%)	225	(11.6%)	303	(19.1%)	136	(14.3%)	9	(0.4%)
5 to 5.9	231	(12.9%)	225	(11.6%)	205	(12.9%)	84	(8.9%)	0	
6 to 6.9	318	(17.7%)	320	(16.5%)	201	(12.7%)	47	(5.0%)	0	
7 to 7.9	284	(15.8%)	271	(13.9%)	86	(5.4%)	18	(1.9%)	0	
8 to 8.9	205	(11.4%)	242	(12.5%)	62	(3.9%)	22	(2.3%)	0	
9 to 9.9	130	(7.2%)	122	(6.3%)	26	(1.6%)	5	(0.5%)	0	
10 to 10.9	98	(5.5%)	71	(3.7%)	11	(0.7%)	3	(0.3%)	0	
11 to 11.9	17	(1.0%)	26	(1.3%)	2	(0.1%)	2	(0.2%)	0	
> 12	21	(1.2%)	33	(1.7%)	6	(0.4%)	0		0	
Total	1794		1943		1586		949		2460	

(Early); 4 to 4.9 kg in cows calved 18 to 22 weeks; and 1 to 1.9 kg in cows calved over 22 weeks (Late).

The percentage of cows with values outside the standard ranges varied greatly between groups of cows, and between metabolites (Table 2.9 and Figure 2.2). One third of cows in the Post-partum group and one quarter in the Early group had raised betahydroxybutyrate. One quarter of cows in the Post-partum group had low glucose. Bile acids were raised in three quarters of cows in the Post-partum, Early and Mid lactation groups, in two thirds of cows in Late lactation, but in only 38% of dry cows.

Urea was above the standard range in over half of cows in each of the groups. High globulins were seen in 15% of cows in Mid lactation, and 22% in dry cows and in Post-partum cows.

Magnesium was below standard in 5% or less of lactating cows, and in under 9% of dry cows. Phosphate was low in around one fifth of cows in all groups.

Glutathione peroxidase was below standard in 18% of cows in the Post-partum group, 2.5% in Mid lactation and 10% in the dry cows.

Table 2.9   NUMBER AND PERCENTAGE OF COWS WITH VALUES OF BLOOD
METABOLITES OUTSIDE STANDARD RANGES

	Post-partum	*Early*	*Mid*	*Late*	*Dry*
BOHB	775	615	280	139	193
< 0.9 mmol/l	34.0%	25.9%	14.8%	10.3%	7.6%
Glucose	563	207	96	91	229
< 2 mmol/l	24.8%	8.8%	5.1%	6.9%	9.1%
Bile acids	1697	1853	1406	848	963
> 60 micromol/l	74.7%	78.5%	74.5%	63.6%	38.1%
Urea	136	85	71	54	181
< 3.3 mmol/l	6.0%	3.6%	3.7%	4.0%	7.1%
Urea	1222	1513	1236	918	1457
> 5 mmol/l	53.6%	64.7%	65.2%	67.9%	57.3%
Albumin	98	64	24	24	50
< 30 g/l	4.3%	2.7%	1.3%	1.8%	2.0%
Albumin	20	81	85	37	24
> 40 g/l	0.8%	3.4%	4.5%	2.7%	0.9%
Globulin	161	146	128	90	215
< 30 g/l	7.1%	6.2%	6.8%	6.7%	8.5%
Globulin	500	476	281	237	566
>40 g/l	22.0%	20.1%	14.9%	17.6%	22.3
Magnesium	90	90	63	67	221
< 0.74 mmol/l	4.1%	3.8%	3.3%	5.0%	8.7%
Phosphate	450	526	399	246	422
< 1.8 mmol/l	19.8%	22.3%	21.1%	18.2%	16.6%
GSHPx	331	162	33	28	197
< 40 units	17.6%	8.8%	2.5%	2.8%	9.8%
Copper	25	33	25	29	79
< 9.4 micromol/l	1.6%	2.1%	1.9%	2.8%	3.2%

BODY CONDITION AT TIME OF SAMPLING,
BETAHYDROXYBUTYRATE, GLUCOSE AND BILE ACIDS

Cows were divided into those below condition score 3 at the time of
sampling ('thin') and those at condition score 3 or above ('fat'). In each
group of cows, mean values of betahydroxybutyrate (BOHB), glucose and
bile acids were calculated for each period of 5 days from 60 days pre-
partum to 325 days post-partum having excluded extreme values (Rowlands, 1984) (Table 2.10 and Figure 2.3).

   BOHB was similar in the thin and fat cows, rising steeply from calving to
a peak at 25 days, and then declining through lactation. Glucose concentration declined in both groups from calving to a minimum between 6 and
20 days, then rose to a peak at 305 days in the thin group and 315 days in the

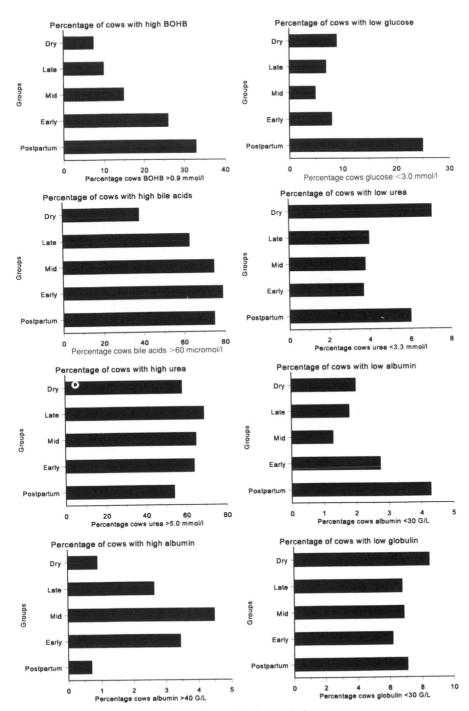

Figure 2.2. Percentage of cows with values outside the standard ranges.

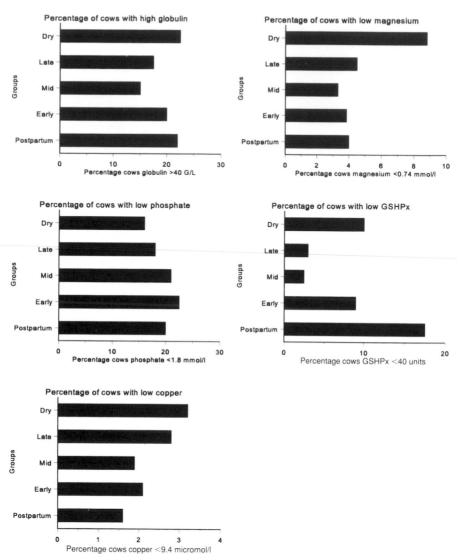

fat group. Cows that were above condition score 3 had a significantly higher glucose than thinner cows between 160 and 200 days (p < 0.01) and after 205 days (p < 0.05). Bile acids rose steeply after calving to a peak at 115 days, then declined until the next calving. Bile acids were significantly higher in the thin cows between 125 and 200 days post-partum (p < 0.001) and after 205 days (p < 0.01).

In the thin group, glucose concentration was negatively correlated with BOHB between 45 and 120 days post-partum (p < 0.01, r − 0.78), but

Table 2.10   BETAHYDROXYBUTYRATE, GLUCOSE AND BILE ACIDS IN COWS WITH CONDITION SCORE AT TIME OF SAMPLING 3 OR ABOVE ('FAT') OR BELOW 3 ('THIN')

Days post-partum	Thin B.hydroxybutyrate mmol/l	Fat B.hydroxybutyrate mmol/l	Thin Glucose mmol/l	Fat Glucose mmol/l	Thin Bile acids micromol/l	Fat Bile acids micromol/l
5	0.57	0.57	3.60	3.45	65.90	58.20
10	0.81	0.80	3.02	3.16	68.60	70.80
15	0.77	0.93	3.17	3.15	69.40	79.20
20	0.85	0.95	3.13	3.05	79.60	79.40
25	0.88	1.04	3.22	3.20	81.60	81.40
30	0.85	0.94	3.27	3.31	84.70	80.00
35	0.81	0.92	3.34	3.37	84.80	81.10
40	0.78	0.98	3.39	3.41	84.40	94.20
45	0.72	0.79	3.42	3.44	83.80	91.40
50	0.79	0.82	3.47	3.49	85.70	82.90
55	0.73	0.80	3.47	3.56	94.60	77.20
60	0.82	0.77	3.43	3.53	89.60	93.50
65	0.70	0.78	3.58	3.57	87.40	81.00
70	0.69	0.66	3.50	3.71	89.10	85.90
75	0.69	0.75	3.57	3.63	82.90	89.30
80	0.71	0.64	3.50	3.71	85.50	79.30
85	0.75	0.74	3.65	3.64	79.70	77.10
90	0.71	0.70	3.61	3.64	93.70	81.50
95	0.68	0.73	3.59	3.67	94.70	90.30
100	0.71	0.73	3.60	3.58	93.30	85.00
105	0.67	0.77	3.59	3.67	75.30	95.50
110	0.65	0.54	3.62	3.69	78.10	78.60
115	0.65	0.66	3.69	3.60	84.30	71.20
120	0.61	0.56	3.67	3.67	79.10	74.60
125	0.63	0.62	3.51	3.77	88.90	79.50
130	0.60	0.61	3.62	3.64	82.60	71.10
135	0.58	0.70	3.68	3.51	91.40	76.80
140	0.58	0.57	3.68	3.73	84.20	70.80
145	0.59	0.67	3.64	3.71	82.70	75.70
150	0.58	0.62	3.55	3.56	82.80	86.70
155	0.63	0.60	3.65	3.67	86.90	81.70
160	0.61	0.65	3.67	3.63	82.60	79.40
165	0.62	0.61	3.72	3.73	86.70	77.60
170	0.63	0.61	3.66	3.60	79.10	86.90
175	0.58	0.58	3.58	3.55	86.10	81.00
180	0.58	0.63	3.60	3.57	80.10	80.10
185	0.55	0.71	3.56	3.59	94.70	82.00
190	0.61	0.72	3.59	3.74	77.30	82.50
195	0.64	0.62	3.65	3.67	77.40	77.80
200	0.72	0.59	3.45	3.77	81.80	69.20
205	0.62	0.68	3.65	3.75	79.90	75.00
210	0.68	0.63	3.65	3.75	91.50	78.50

Table 2.10   *Continued*

Days post-partum	Thin B.hydroxybutyrate mmol/l	Fat B.hydroxybutyrate mmol/l	Thin Glucose mmol/l	Fat Glucose mmol/l	Thin Bile acids micromol/l	Fat Bile acids micromol/l
215	0.61	0.63	3.54	3.60	75.90	79.40
220	0.62	0.60	3.73	3.84	85.40	68.50
225	0.54	0.60	3.66	3.64	84.90	67.70
230	0.55	0.52	3.70	3.77	78.50	73.40
235	0.60	0.54	3.69	3.77	77.90	81.60
240	0.61	0.62	3.55	3.78	75.60	72.80
245	0.63	0.65	3.40	3.80	80.30	71.50
250	0.50	0.49	3.57	3.59	79.60	73.00
255	0.55	0.54	3.79	3.69	77.70	69.60
260	0.47	0.59	3.68	3.62	79.10	68.10
265	0.49	0.52	3.63	3.70	74.30	65.30
270	0.56	0.61	3.67	3.71	75.60	73.30
275	0.60	0.56	3.66	3.77	70.90	63.10
280	0.66	0.57	3.47	3.70	57.10	57.40
285	0.52	0.48	3.56	3.63	67.90	65.40
290	0.49	0.53	3.32	3.60	54.40	52.20
295	0.57	0.49	3.63	3.71	75.80	59.10
300	0.47	0.53	3.36	3.73	55.10	64.00
305	0.57	0.52	3.83	3.59	87.40	68.70
310		0.43		3.60		71.60
315		0.63		3.77		76.60
320		0.44		3.76		64.00
325		0.57		3.57		56.90
-60	0.57	0.53	3.60	3.76	76.00	59.80
-55	0.59	0.54	3.61	3.67	68.10	55.50
-50	0.50	0.54	3.46	3.61	63.40	55.90
-45	0.44	0.51	3.38	3.57	62.40	54.40
-40	0.57	0.56	3.64	3.59	64.90	52.80
-35	0.53	0.53	3.29	3.58	59.00	52.10
-30	0.57	0.54	3.58	3.59	61.30	53.60
-25	0.58	0.56	3.45	3.50	60.30	53.60
-20						
-15	0.54	0.52	3.52	3.55	57.20	52.50
-10	0.57	0.55	3.40	3.48	55.30	52.30
-5	0.55	0.51	3.43	3.52	58.20	56.80
-1	0.51	0.53	3.51	3.45	51.90	59.00

positively correlated with BOHB after 205 days ($p < 0.05$, r 0.54), and with bile acids from 125 to 200 days ($p < 0.02$-0.02, r 0.42) and after 205 days ($p < 0.01$, r 0.79). In fat cows, glucose was again negatively correlated with BOHB ($p < 0.01$, r − 0.75), and bile acids correlated with BOHB from 125 to 200 days ($p < 0.01$, r 0.52).

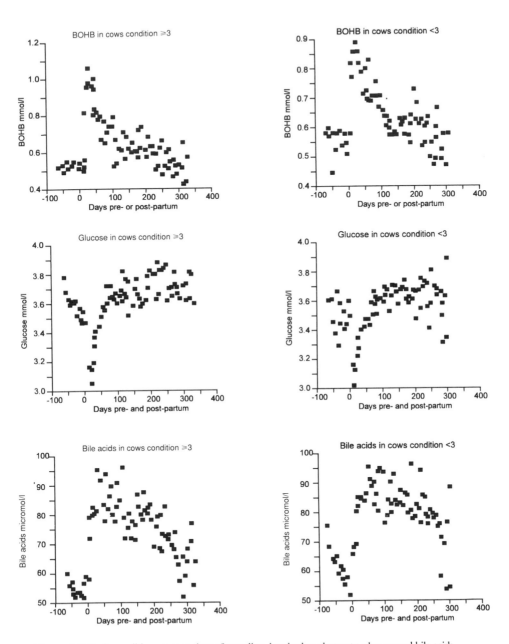

Figure 2.3. Body condition score at time of sampling, betahydroxybutyrate, glucose and bile acids.

CONDITION SCORE, BETAHYDROXYBUTYRATE, GLUCOSE AND BILE ACIDS IN COWS SAMPLED TWICE

A total of 430 cows were found to have been tested both during the dry period and subsequently in early lactation (up to 17 weeks post-partum). A total of 73 were above condition score 3 ('fat') both when dry, and when sampled after calving: 174 were above condition score 3 when dry, but below ('thin') after calving, and 183 were below condition score 3 before and after calving. Very few cows moved from 'thin' to 'fat' after calving.

For each 5-day period, mean values of betahydroxybutyrate, glucose and bile acids were calculated, between 1 and 16 results contributing to each mean (Figure 2.4). In cows sampled before 40 days, betahydroxybutyrate and bile acids were higher and glucose was lower ($p < 0.05$) in cows fat before calving than in those cows thin before calving.

GLUTATHIONE PEROXIDASE AND LACTATION NUMBER

Glutathione peroxidase concentration was below standard values in over 17% of first lactation cows (heifers), whereas low values were seen in no more than 9% of mature cows (Table 2.11). During the Post-partum period in first lactation cows, 39% had low GSHPx values.

Discussion

Metabolites were chosen which fluctuate in a meaningful way in dairy cows. Calcium, for example was excluded because homeostatic mechanisms usually maintain the concentration within a close range except in cows with signs of clinical disease. Metabolites were also chosen where changes were likely to indicate a problem about which advice could usefully be given. Packed cell volume (PCV) for example, fluctuates in an interesting fashion during the year but the information is of limited value. Metabolites that could not withstand storage for 24 or more hours at ambient temperatures were excluded.

Bile acids were chosen to indicate liver function: Roberts, Red, Rowlands and Patterson (1981) at Compton showed the widespread existence and clinical importance of fat mobilisation syndrome, and West (1990, 1991) showed the close correlation in cattle between fatty infiltration of the liver and the concentration of bile acids in the blood.

The minerals chosen were those believed to be most frequently involved in subclinical problems, affecting productivity in British dairy cattle.

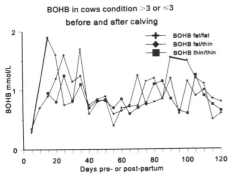

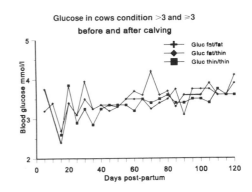

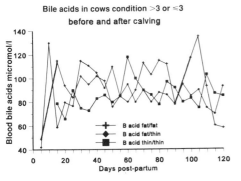

Figure 2.4. Body condition score, betahydroxybutyrate, glucose and bile acids in cows sampled twice.

The standard ranges must be to some extent arbitrary, but are based on the best available information on the laboratory tests used, and the population of cows tested.

OVERVIEW

One aim of the DHMS was to correlate information about the herd and the cows with the blood results. The information has been reasonably com-

Table 2.11 NUMBER AND PERCENTAGE OF COWS IN EACH LACTATION WITH GLUTATHIONE PEROXIDASE BELOW STANDARD VALUES (40 UNITS)

Lactation	1*	2	3	4	5	6	> 6
Number < 40 units	218	151	121	81	85	62	103
Total number	1254	1836	1687	1361	1042	777	1139
Percentage	17.4	8.2	7.2	6.0	8.2	8.0	9.0

*In first lactation between calving and 6 weeks post-partum, 39.1% of cows had glutathione peroxidase below 40 units.

plete, but in the future it is planned to ensure more comprehensive collection of information.

The aim of encouraging veterinarians to discuss nutrition with farmers and nutritionists has been largely successful. Veterinarians have invariably been involved in the collection of the blood, and have received reports, which always refer to matters which the farmer is advised to discuss with the veterinarian and nutritionist. A survey is currently under way to assess the perceptions of the veterinarians to the DHMS.

The selection of farms is not random. It is inevitable that the distribution of farms will reflect the enthusiasm of farmers and of cattle specialists and representatives. Nor will the distribution of cows sampled be random: farmers may be tempted to present the 'best cow' for sampling, in the belief that the blood sample will reveal her secret, or they may keep in a cow with a problem, thinking that the metabolic profile will explain it. Both of these practices were discouraged, but if they have occurred, they would be expected to extend the range of some values.

Most of the farms have been in the North and Midlands of England, with 22 from Scotland. The geographical area is limited, but the wide range of husbandry patterns and climate are representative of much of Great Britain.

The numbers of cows in each lactation and the numbers of cows calving in each month produce patterns similar to those in the area as a whole. Milk yields show a pattern that would be common in many British herds, with 44% of those calved 6 to 18 weeks (Early lactation group) giving between 30 and 39 kg a day, 40% yielding 20 to 29 kg, and 7% above 40 kg.

The pattern of condition score is interesting, with 57% of dry cows above condition score 3 at the time of sampling, and only 5% below condition score 2.5. Since 23% of cows sampled up to 6 weeks after calving (Post-partum group) were below condition score 2.5, it appears that some cows were losing a considerable amount of body condition in a short time. At no stage were more than 9% of cows below condition score 2. There are clearly a lot of cows too fat in the dry period, and they are likely to eat less after calving than leaner cows, to lose more condition and to have a lower fertility (Garnsworthy and Topps, 1982; Garnsworthy, 1988; Treacher, Reid and Roberts, 1986).

Parlour feed shows a wide range in each category of cows. For cows calved 6 to 18 weeks (Early group), the largest number are receiving 6 to 7 kg a day, while in mid lactation the greatest number are on 4 to 5 kg. Despite the trend to feed more straights, and to reduce dependence on parlour feed, over 25% of cows in the Early group were receiving more than 8 kg of parlour feed a day.

ABNORMAL VALUES

It is no surprise that BOHB is high in one third of cows up to 41 days post-partum, and in a quarter of cows between 42 and 120 days. A single individual value above the normal range is less significant than several raised values in the same group of cows, for two reasons. First, a cow may have an individual problem unrelated to the rest of the herd. Secondly, the high degree of variation found in BOHB (Manston *et al.*, 1981) must mean that a single high value may not be representative. The finding of several high values in a group, or the finding of a very high value, is however interpreted to mean that there is a significant energy deficit. The farmer is advised to discuss with the veterinarian and nutritionist possible causes, such as low palatability of silage, poor access to silage, etc.

Glucose values below 2.5 mmol/l have been regarded as below normal, but in view of the patterns seen, a cut-off point of 3.0 has been adopted for this review. A quarter of cows calved less than 42 days have low glucose, similar to the findings of Rowlands, Manston, Stark, Russell, Collis and Collis (1980) whereas after that time no more than 9% do so, despite the fact that 26% have high BOHB. Cows with high BOHB and normal glucose are commonly seen, and can be explained as cows that are mobilising fat as an energy source. It is less usual to see cows with normal BOHB and depressed glucose. Only occasionally are these thin cows, with little fat to mobilise.

Bile acids are above 60 micromol/l in a large majority of lactating cows. A maximal value of 45 micromol/l was established by West (1991). It is accepted that a high proportion of British dairy cows under current husbandry have fatty infiltration of the liver (Roberts *et al.*, 1981) and high bile acids reflect that (West, 1990, 1991). Since 57% of the dry cows in the herds studied were above body condition 3.0, these would be expected to be at risk from fat mobilisation syndrome (Garnsworthy and Topps, 1982; Roberts *et al.*, 1981). Bile acids show considerable variation in time (Pearson *et al.*, 1992) and single cows with elevated values are not regarded as significant. In many herds, however, high or very high bile acid values were seen in the majority of lactating cows.

High urea is clearly very common in these herds. While numerous studies have been published (eg. Lewis, 1957; Ropstad, Vik-Mo and Refsdal, 1989) there is room for a new study to assess urea values in cows fed according to the new Metabolisable Protein system (Agriculture and Food Research Council, 1989). Further studies would be of value to assess the relationship of blood urea concentration to fertility and health, as well as to production.

Albumin is regarded as another indicator of liver function, and low values can be seen for example in liver fluke infestation. This is a rare occurrence in this population of cows. Albumin and globulin are regarded as indicators of long-term protein intake, and inadequacy was rare in these herds. High globulin can reflect various types of infection, such as chronic mastitis, when the concentration of gamma-globulin may be raised, and high globulin values were found in about one fifth of cows.

Magnesium varies little in time, and low values are regarded as significant. Very little is stored in the body. On occasion very low values were seen, and the veterinarian was then telephoned immediately so that the farmer could be warned. For dry cows, where almost 9% of cows showed low values, low magnesium is a risk factor for milk fever (Sansom, Manston and Vagg, 1983).

Low phosphate was seen in 18 to 22% of lactating cows, most of which were receiving mineralised compound, but below 17% of dry cows, presumably reflecting the higher needs of lactation. In most cases samples were taken from the tail (coccygeal) blood vessels, but when samples were taken from the jugular vein, a standard value of more than 1.6 mmol/l was adopted to allow for the lower concentration in the jugular, where some phosphate is removed by the salivary glands.

Glutathione peroxidase concentration was calculated assuming a constant packed cell volume of 33%. Our own tests agree with Payne *et al.* (1970) that packed cell volume varies between cows and with seasons. Standard values for glutathione peroxidase differ between laboratories, and this causes confusion. Since glutathione peroxidase is incorporated into the red blood cells, the concentration reflects the amount of selenium available to the cow 4 to 6 weeks earlier. This was demonstrated in the low values seen in the dry cows, reflecting low selenium intake in late lactation, and still lower values seen in the Post-partum group, reflecting lack of mineral supplementation in many dry cows. The reports distinguish between marginally low glutathione peroxidase (20 to 40 units), and very low values (below 20 units). Selenium-responsive conditions are by no means seen on farms where cows have marginally low values. The greater use of unmineralised straights in place of mineralised compounds, and the more widespread feeding of maize silage in place of grass silage, may increase the prevalence of selenium-responsive problems.

Copper concentration 9.4 micromol/l and above was adopted as standard (Suttle, 1993) for plasma. When on rare occasions serum was analysed, 7.4 micromol/l was used as standard. Low copper concentration was not found to be common, but two reservations must be expressed. First, blood copper does not always directly relate to clinical problems (Suttle, 1986). Secondly, interference by molybdenum and other elements can cause clinical problems, and blood concentration of copper is not a sufficient guide.

BODY CONDITION AT TIME OF SAMPLING, BETAHYDROXYBUTYRATE, GLUCOSE AND BILE ACIDS

The fall in glucose after calving agrees with Rowlands *et al.* (1980) and the negative correlation with betahydroxybutyrate (BOHB) agrees with Mills, Beitz and Young (1986).The enormous need of the lactating mammary gland for glucose in order to produce lactose means that some energy is produced from mobilised fat, and BOHB is produced.

In general the patterns of these three metabolites were similar in the fat and thin cows. The distinction between the cows was solely the condition at the time of sampling, and it is likely that the change in condition prior to sampling is more important. Further analysis of data from the much smaller number of cows that by chance were sampled twice has therefore been undertaken.

CONDITION SCORE, BETAHYDROXYBUTYRATE, GLUCOSE AND BILE ACIDS IN COWS SAMPLED TWICE

In cows sampled twice, first in the dry period and then before 40 days after calving, betahydroxybutyrate and bile acids were higher when the cows were above condition score 3 before calving than in cows that were leaner. Glucose values, and all values after 40 to 50 days appear to show no difference. This is compatible with evidence that cows calving above condition score 3 mobilise more body tissues and are more likely to have fatty liver (Garnsworthy, 1988; Roberts *et al.*, 1981).

GLUTATHIONE PEROXIDASE AND LACTATION NUMBER

In cows (heifers) in first lactation a low glutathione peroxidase concentration was seen in 17% of cases. An even higher proportion of first lactation cows in the Post-partum group showed low GSHPx (39%). This is a strong indication of a low selenium content in the forage fed to the heifers before their first calving, and an absence of supplementary minerals. The scheme did not set out to analyse blood from heifers before their first calving, but they could well benefit from such attention.

NORMAL VALUES OR MEAN VALUES?

Unlike Payne *et al.*, (1970) normal values have not been calculated as the mean of the values found in apparently normal cows. Instead, standard values have been decided upon from published data. On that basis, the mean values in some groups of cows that have been tested are in many cases not normal. The results reported here suggest that a large number of British dairy cows are too fat in the dry period, lose too much condition in early lactation, and have energy deficit and fatty infiltration of the liver. An excess of urea, indicating excess rumen-degradable protein relative to fermentable energy, is very common. Low magnesium in dry cows is not uncommon, low phosphate is commonly seen, and low glutathione peroxidase is very common early in first lactation.

If mean values were taken as normal, and values more than two standard deviations from the mean as abnormal, the results of any common fault in husbandry would become accepted as normal.

FUTURE ADVANCES IN METABOLIC PROFILES

Until now, we have encouraged the use of blood samples, and a beneficial effect of that has been that it necessarily involves the veterinarian. Milk samples can, however, be used for the measurement of urea, and the concentration is closely correlated with blood concentration (Oltner and Wiktorsson, 1983; Ropstad *et al.*, 1989; Gustafsson and Palmquist, 1993). Milk samples have for many years been used to measure total ketones, and the values are less variable than in urine because the concentration of urine fluctuates widely (Radostits, Blood and Gay, 1994). Other metabolites such as glucose would clearly bear no relationship to blood values. Standard ranges would need to be established for suitable metabolites. Milk has the enormous advantage that it is collected twice daily and bulk samples are regularly taken to central laboratories, and could be analysed without the expense of collection. Results could be interpreted alongside the milk protein and butterfat values, which are already routinely measured, and which are useful indicators of nutritional status. Individual samples could be collected easily by the farmer. Frequent measurement of urea could be used as a means of monitoring the balance of nitrogen and energy available to the rumen microbes.

At present, standard values are used without formal regard for the age or lactational status of the individual cow. In future, computer software could highlight values outside standard ranges depending on age, days after calving, time of year, etc.

Bile acids have been used as the best available single predictor of fatty liver (West, 1991) but it is likely that other metabolites might add to the value of the prediction. Indicators of other functions of the liver would also be of value.

Questionnaires have recently been circulated to cattle specialists and representatives, farmers and veterinarians who have been involved in the scheme. New developments in presentation and in substance will take account of the opinions expressed.

Other companies have shown an interest in metabolic profiles. The particular package developed with AF plc has not been offered elsewhere, but collaborative work with other organisations has been undertaken. It is clearly beneficial for experience to be shared, though of course confidentiality has to be respected.

Preliminary trials have been made of metabolic profiles in beef cattle and in sheep, which clearly have much in common with each other and with dairy cows. Trials have also been undertaken in pigs, where the range of useful metabolites will probably be very different: the greater variability in the diet of ruminants may mean that metabolic profiles are peculiarly valuable. Heifers have not been included in the scheme, but the evidence of low selenium in the diets of many heifers before calving suggests that extension of the scheme to heifers would be useful.

A feature of the Dairy Herd Monitoring Scheme has been to link the blood values to other information. More complete and more standardised recording of health and fertility, as in the second DAISY report (Esslemont and Spincer, 1994) could add to the value of the whole scheme.

Conclusions

The University of Liverpool Veterinary Faculty, with AF plc, have shown that metabolic profiles have a place in modern dairy farming. All the parties involved have benefited from a better understanding of the complex links between nutrition, health and fertility. These benefits arise to a large extent through the collection of information about the herd and the individual cows, and discussion of this information alongside the laboratory results.

Acknowledgments

Laboratory work has been efficiently and meticulously performed by Malcolm Savage, Andrew Wattrett and Cynthia Dare, of the Department of

Veterinary Pathology, University of Liverpool, Leahurst. Secretarial work has been efficiently done by Christine Broadbent.

References

Agriculture and Food Research Council (1989) Technical Committee on Responses to Nutrients, Report No. 9, Nutritive Requirements of Ruminant Animals: Protein, *Nutrition Abstracts and Reviews*, Series B, Livestock Feeds and Feeding, **62**, 787–835

Anderson, P.H., Berrett, S. and Patterson, D.S.P. (1978) Glutathione peroxidase activity in erythrocytes and muscle of cattle and sheep and its relationship to selenium. *Journal of Comparative Pathology*, **88**, 181–189

Esslemont, R.J. and Spincer, I. (1994) *DAISY Report Number 2*. Reading: University of Reading

Garnsworthy, P.C. (1988) The effect of energy reserves at calving on performance of dairy cows. In *Nutrition and lactation in the dairy cow*. Edited by P.C. Garnsworthy, pp. 157–170. London: Butterworth

Garnsworthy, P.C. and Topps, J.H. (1982) The effect of body condition of dairy cows at calving on their food intake and performance when given complete diets. *Animal Production*, **35**, 113–119

Gustafsson, A.H. and Palmquist, D.L. (1993) Diurnal variation of rumen ammonia, serum urea, and milk urea in dairy cows at high and low yields. *Journal of Dairy Science*, **76**, 475–484

Lewis, D. (1957) Blood urea concentration in relation to protein utilisation in the ruminant. *Journal of Agricultural Science*, **48**, 438–446

Manston, R., Rowlands, G.J., Little, W. and Collis, K.A., (1981) Variability in the blood composition of dairy cows in relation to time of day. *Journal of Agricultural Science*, **96**, 593–598

Mills, S.E., Beitz D.C. and Young, J.W. (1986) Characterisation of metabolic changes during a protocol for inducing lactation ketosis in dairy cows. *Journal of Dairy Science*, **69**, 352–361

Oltner, R. and Wiktorsson, H. (1983) Urea concentrations in milk and blood as influenced by feeding varying amounts of protein and energy to dairy cows. *Livestock Production Science*, **10**, 457–467

Parker, B.N.J. and Blowey, R.W. (1976) Investigations into the relationship of selected blood components to nutrition and fertility of the dairy cow under commercial farm conditions. *Veterinary Record*, **98**, 394–404

Payne, J.M., Dew, S.M., Manston, R. and Faulks, R. (1970) The use of a metabolic profile test in dairy herds. *Veterinary Record*, **87**, 150–158

Pearson, E.G., Craig, A.M. and Rowe, K. (1992) Variability of serum bile acid concentrations over time in dairy cattle, and effect of feed deprivation on the variability. *American Journal of Veterinary Research,* **53,** 1780–1783

Radostits, O.M., Blood, D.C. and Gay, C.C. (1994) *Veterinary Medicine,* 8th edition, p. 1348. London: Bailliere Tindall

Roberts, C.J., Reid, I.M., Rowlands, G.J., Patterson (1981) A fat mobilisation syndrome in dairy cows in early lactation. *Veterinary Record,* **108,** 7–9

Ropstad, E., Vik-Mo, L, and Refsdal, A. O. (1989) Levels of milk urea, plasma constituents, and rumen liquid ammonia in relation to the feeding of dairy cows during early lactation. *Acta veterinaria scandinavica,* **30,** 199–208

Rowlands, G.J., Manston, R., Stark A.J., Russell A.M., Collis, K.A., and Collis, S.C. (1980) Changes in albumin, globulin, glucose and cholesterol concentrations in the blood of dairy cows in late pregnancy and early lactation and relationships with subsequent fertility. *Journal of Agricultural Science,* **94,** 517–527

Rowlands, G.J. (1984) Week-to-week variation in blood composition of dairy cows and its effect on interpretation of metabolic profile tests. *British Veterinary Journal,* **140,** 550–557

Sansom, B.F., Manston, R. and Vagg, M.J. (1983). Magnesium and milk fever. *Veterinary Record,* **112,** 447–449

Suttle, N. (1986) Problems in the diagnosis and anticipation of trace element deficiencies in grazing livestock. *Veterinary Record,* **119,** 148–152

Suttle, N. (1993) Overestimation of copper deficiency. *Veterinary Record,* **133,** 123–124

Treacher, R.J., Reid, I.M. and Roberts, C.J. (1986) Effect of body condition at calving on the health and performance of dairy cows. *Animal Production,* **43,** 1–6

West H.J. (1990) Effect on liver function of acetonaemia and the fat cow syndrome in cattle. *Research in Veterinary Science,* **51,** 133–140

West H.J. (1991) Evaluation of total serum bile acid concentrations for the diagnosis of hepatobiliary disease in cattle. *Research in Veterinary Science,* **48,** 221–227

3

FEEDING DAIRY COWS OF HIGH GENETIC MERIT

L.E. CHASE

Department of Animal Science, Cornell University, Ithaca, NY 14853, USA

Introduction

The genetic base for milk production continues to increase in both the individual cow and total herd milk production. The current world record milk production for an individual cow is in excess of 26,700 kg. The highest producing herd in the US has a herd average of 14,700 kg/cow/year from 440 milking cows. Average herd milk production for US Holstein herds on Dairy Herd Improvement recording programmes increased from 4,160 to 8,178 kg/cow/year from 1950 to 1990. What will herd averages be 10 years from now?

A challenge exists in developing feeding and management programmes for these herds. The herd manager expects and demands nutrition programmes that control the feed cost per unit of milk while supporting high levels of milk production. At the same time, herd health and reproductive performance must be maintained. Milk components must also be considered, based on the formulae used to determine the price of milk. Milk pricing systems in the US are shifting from milk-fat based to either a protein or nonfat solids basis. These shifts in pricing structure will alter ration formulation strategies.

One challenge is to define the appropriate requirements to use in formulating rations for high producing dairy cows. A number of nutrient requirement systems are used throughout the world (AFRC, 1993; NRC, 1989), based on research data where available. The majority of the data used as a base for developing these requirements were obtained using cows producing less than 9,000 kg of milk per lactation. Can these requirements be used when designing rations for high producing herds?

Nutrient utilization

Nutrients consumed by the dairy cow must be partitioned to meet demands for maintenance, growth, reproduction and milk production. These requirements will be met in a priority order by the cow depending upon the quantity of nutrients available. A mature cow will have maintenance as the highest priority followed by milk production and then reproduction. As dry matter intake (DMI) increases, a smaller proportion of the total nutrient intake is used for maintenance. Figure 3.1 shows the proportion of the net energy (NE) and crude protein (CP) available to support milk production levels. The proportions of the protein and energy intake available for milk production are 73 and 49% for a cow producing 15 kg of milk. These figures increase to 89 and 74% in a cow producing 45 kg of milk.

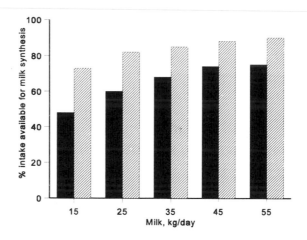

Figure 3.1 Percentage of total nutrient intake used for milk production.

Dry Matter Intake

The single most important variable which influences productivity in dairy cows is DMI (Mertens, 1987; Waldo, 1986). Surveys of high producing herds in England and Wales indicated that attaining high levels of DMI was a key management attribute (Neilson, Whittemore, Lewis, Alliston, Roberts, Hodgson-Jones, Mills, Parkinson and Prescott., 1983; Wood and Wilson, 1983). There have been a number of equations proposed to predict DMI of dairy cows (Roseler, Fox, Chase and Stone, 1993; NRC, 1989). However, none of these have been validated with data from high producing dairy cows. An additional consideration is the rate of increase in DMI in early lactation. A depression of 18% from predicted DMI is indicated for early lactation cows (NRC, 1989) and weekly adjustment factors for DMI have been developed (Kertz, 1991; Roseler *et al.*, 1993). In the first week postcalving, DMI is about 67% of the maximum attained at 8 weeks into lactation. This depressed DMI in early lactation presents a challenge in terms of meeting energy needs while maintaining fibre intake and rumen function.

The energy value of feeds is also altered by DMI. A positive relationship exists between DMI and rate of passage of feed through the cow (Colucci, Chase and Van Soest, 1982; Mertens, 1987). This depression in digestibility with increasing DMI is not uniform for all feeds and feed fractions. Feed composition tables containing discounts for individual feeds are available (Van Soest and Fox, 1992). Computer modelling approaches may be needed to make these adjustments in a more systematic and practical manner (Sniffen, O'Connor, Van Soest, Fox and Russell, 1992). Current nutrient requirement publications do provide for adjustments in feed energy values based on level of intake (AFRC, 1993; NRC, 1989).

Forage quality

Forages provide the base for designing dairy rations. Potential forage intake is at least partially related to neutral-detergent fibre (NDF) content (Kawas, Jorgensen and Danelon, 1991; Waldo, 1986). Kawas *et al.* (1991) indicated that dietary concentrations should be less than 20% acid-detergent fibre (ADF) or 29% NDF for cows between weeks 10 and 26 of lactation. The cows in this study were producing more than 35 kg of 4% fat-corrected milk (FCM). The relationship between forage quality, concentrate supplement-ation and milk production has been examined with alfalfa forages (Kaiser and Combs, 1989; Kawas *et al.*, 1991; Llamas-Lamas and Combs, 1990 and 1991). Fibre digestion was more rapid in early vegetative hay than in either late bud or full bloom hay (Llamas-Lamas and Combs, 1990). No differ-ences were detected in milk production when rations contained similar dietary fibre contents with three qualities of alfalfa and varying forage to concentrate ratios (Kaiser and Combs, 1989). Similar levels of milk produc-tion were reported in mid-lactation dairy cows fed rations with either high quality orchard-grass or alfalfa silages (Weiss and Shockey, 1991). There are differences in the composition and digestion of the NDF fractions of legume and grass forages. Total NDF is higher in legumes than grasses. However, the cell-soluble fraction is lower in grasses. Ruminal and total tract digestibilities for both ADF and NDF were higher in rations contain-ing orchard grass than those containing alfalfa (Holden *et al.*, 1994). A forage source with high digestibility and intake is still the key to development of rations to support high levels of milk production.

Carbohydrate and protein

Nutrient utilization and microbial protein synthesis in the rumen are the keys to designing efficient and profitable dairy rations. A number of review articles have examined the protein and carbohydrate considerations in optimizing rumen fermentation (Clark, Klusmeyer and Cameron, 1992; Chalupa and Sniffen, 1994; Hoover and Stokes, 1991; Nocek, 1994).

The carbohydrate components of feeds can be divided into structural and nonstructural fractions. The structural component includes pectin, cellulose, hemicellulose and lignin. The ADF fraction consists of cellulose plus lignin. The ADF plus hemicellulose comprises NDF. The nonfibrous carbohydrate (NFC) fraction is calculated as follows:

$$NFC = 100 - (\%\,CP + \%\,NDF - \%\,Ash + \%\,EE)$$

where EE = ether extract.

This fraction is also termed nonstructural carbohydrate (NSC) in some publications. The main components of this fraction are sugars, starches, pectin, galactans and beta-glucans. In fermented feeds, volatile fatty acids and lactic acid are in this fraction. The calculated NFC values should be adjusted for the presence of the organic acids (AFRC, 1993). More information is needed on the rates of digestion of the carbohydrate fractions. Depending upon the proportion of sugars and starches, degradation rates would be expected to vary significantly. Currently published digestion rates are based on very limited data (Sniffen *et al.*, 1992). There are also digestion rate differences between feed types. *In vivo* ruminal starch degradation values for wheat, oats and barley are about 84–90% of the total starch. Similar values for corn and sorghum grain were 67–76%. Differences also exist due to processing method and particle size. Relative rates of degradation (fastest to slowest) are steam-flaked > high moisture > dry ground > dry rolled > dry whole for corn and sorghum grains.

The protein components of feeds are divided into rumen degradable (DIP) and rumen undegradable (UIP) in the new dairy requirement system (NRC, 1989). In some sectors, a soluble protein (SIP) fraction is also used. This is the rapidly available portion of DIP as determined in a borate-phosphate buffer with a 1 hour incubation. This subdivision of protein fractions is similar in concept to the AFRC (1993) system. Amino acid nutrition is an integral component of the protein utilization system. This area has been recently reviewed (Chalupa and Sniffen, 1994; Schwab, 1994). The available knowledge in this area continues to grow. However, there are still significant questions relative to amino acid nutrition of dairy cattle which require additional research. The amount of microbial protein synthesized in the rumen and its contribution to the amino acid supply requires better quantification. Additional information on the amino acid content of the UIP fraction of feeds is also needed. A recent paper determined the amino acid profiles in the original feedstuff and the UIP fraction (Erasmus *et al.*, 1994). The lysine concentration in the UIP fraction was lower than in the original feed for 9 out of 12 feeds. The intestinal digestibility of the UIP fraction ranged from 56% for blood meal to 98% for soya bean meal. This variation in intestinal digestibility of the UIP fraction has been previously reported in other papers. It appears that the amino acid balance of dairy cows can be estimated using computer models (Sniffen *et al.*, 1992; Chalupa and Sniffen, 1994; Chase, 1991).

The relationship between carbohydrates, proteins, rumen fermentation and milk production has been reviewed (Hoover and Stokes, 1991). The concept of matching the rates of ruminally available carbohydrate and protein should enhance rumen fermentation, microbial protein synthesis (MPS) and animal performance. Nocek (1994) suggested that rations for high producing dairy cows should contain 78% total carbohydrate and 53% ruminally available carbohydrate. Hoover and Stokes (1991) examined the relationships of NSC and DIP needed to optimize rumen microbial yield. In their studies, NSC level as a percentage of total carbohydrates was held constant at 56%. Maximum MPS was attained when rations contained 10-13% DIP on a dry matter basis. Additional research work is needed to examine the impact of NSC subfractions on MPS and milk production.

One challenge in implementing the protein systems proposed by both NRC (1989) and AFRC (1993) is defining the proportion of the total protein in the various fractions. Tabular values are available (NRC, 1989; AFRC, 1993; Chalupa and Sniffen, 1994). However, many of these values are based on limited data. The published tabular values also do not provide an index of the variability which exists. The mean protein degradability for soya bean meal analysed in 23 laboratories was 63% with a standard deviation of 11.1% (Madsen and Hvelplund, 1994). The variation reported in this paper is primarily laboratory variation. Stern, Calsmiglia and Endres (1994a) reported a UIP value of 25% for soya bean meal with a standard deviation of 3%. This value was for 5 samples of soya bean meal evaluated by the *in situ* method within the same laboratory.

A second area of ongoing research is to describe better the effect of processing methods on ruminal protein degradation. Heat or chemical treatments will decrease the protein degradability of soya bean meal in the rumen (Stern, Varga, Clarke, Firkins, Huber and Palmquist, 1994b). Heat treating of cereal grains has also been reported to significantly decrease ruminal protein degradability (McNiven *et al.*, 1994). Extensive work has also been reported on the effect of various processing times and temperatures on the UIP value of whole soya beans (Faldet, Satter and Broderick, 1992; Satter, Dhiman and Hsu, 1994). These treatment methods should enhance our ability to control the nitrogen use between the rumen and postruminal segments of the digestive tract.

It is also accepted that the rumen degradability of protein sources varies with outflow rate. The effective rumen degradability of protein sources tends to decrease as rumen outflow rates increase. Protein degradability of soya bean meal was reported to be 86.9, 74.2 and 66% at fractional outflow rates of 0.02, 0.05 and 0.08 per hour (Erasmus *et al.*, 1988). The protein fraction values listed in AFRC (1993) reflect this situation.

The changes in protein fractions of forages which occur during harvest and storage are important factors in formulating rations. Their importance increases as dairy herds utilize a higher proportion of silages in the ration. During both the wilting and fermentation processes, some of the true protein is converted to soluble protein and non-protein nitrogen (NPN). Field observations indicate that soluble protein may comprise 50-80% of the total protein in silages. The degradation rate of both total and chloroplast membrane proteins were higher for silages than the original, fresh forage (Makoni *et al.*, 1994). Nitrogen metabolism was investigated in dairy cows fed direct-cut or wilted grass silages (Teller *et al.*, 1992). Organic matter intake was higher for cows fed the wilted grass silage. Nitrogen and amino acid flow to the duodenum was also higher for cows fed wilted silage. An 18% increase in milk production was reported when an UIP source was added to alfalfa based rations (Dhiman *et al.*, 1993). This implies a shortage of intestinally available protein and amino acids even though the base ration contained an excess of total protein.

High producing herds

Field observations of feeding programmes in high producing dairy herds can be useful in examining how nutritional concepts are put into practice. This information must be interpreted cautiously since DMI is estimated rather than measured as in a research context.

A survey of 61 high producing herds in the US was conducted in 1991

Table 3.1 MILK PRODUCTION AND RATION NUTRITIONAL PARAMETERS FOR SELECTED HOLSTEIN HERDS[a]

Item	Herd								Average	NRC (1989)[c]
	1	2	3	4	5	6	7	8		
Milk, Kg/cow	14,167	13,590	13,353	13,103	12,818	13,070	12,260	11,740	13,012	–
Milk fat, %	3.7	3.8	3.5	3.7	3.6	3.4	3.7	3.9	3.7	–
Milk protein, %	3.2	3.4	3.2	3.1	3.0	3.0	3.1	3.1	3.1	–
DMI, kg	26.7	24.0	26.3	25.5	24.7	22.9	23.6	22.3	24.5	–
Ration DM %	42.5	60.0	–	58.0	60.0	56.0	67.0	58.0	57.4	–
CP, %	18.5	18.0	20.0	18.6	19.6	20.0	19.4	19.0	19.1	18
UIP, % of CP	38.0	41.0	34.0	37.3	37.0	37.3	36.2	34.0	36.8	35
SIP, % of CP	29.0	31.0	32.0	–	–	–	–	–	–	–
NE_1, MCal/kg	1.76	1.76	1.76	1.72	1.80	1.76	1.76	1.69	1.75	1.72[b]
Fat, %	6.9	5.0	5.5	5.1	8.8	7.5	7.3	6.7	6.6	3(min)
ADF, %	19.1	20.0	17.0	19.2	19.9	19.5	20.4	19.1	19.3	19(min)
NDF, %	31.2	32.0	27.0	29.0	29.2	29.0	28.6	28.7	29.3	25(min)
NDF-Forage, %	22.1	20.0	19.0	22.1	16.1	23.0	21.6	20.6	20.6	–
NFC, %	35.0	38.0	37.0	39.6	34.0	34.8	36.3	35.8	36.3	–

[a] Adapted from Chase, 1993 and Howard and Shaver, 1991
[b] Equivalent to about 7.3 MJ/kg of NE_1
[c] NRC (1989) guidelines for 600 kg cows producing 50 kg of milk with 4% milk fat

(Jordan and Fourdraine, 1993). These herds averaged 230.7 milking cows. Average annual milk production was 11,096 kg with 394.5 kg of fat and 347.1 kg of protein. In this survey, 67.2% of the producers used a total mixed ration (TMR) feeding system. Corn silage was fed to 67.2% of the milking herds. Legume hay or haylage was fed to about half of the herds while grass forage was fed to 24.5% of the milking herds. There were also a large number of commodity feeds fed in these herds. The most common ones (with percentage of total herds in brackets) were whole cottonseed (72.1%), distillers grains (37.7%), blood meal (37.7%), meat and bone meal (34.4%) and fish meal (19.7%). Feed additives used were sodium bicarbonate (75.4%), magnesium oxide (67.6%), yeast (50.8%) and niacin (37.7%). Even though this survey provides useful information, it does not permit a detailed analysis of the feeding programme.

Detailed information on feeding programmes was obtained in 13 high producing Holstein herds (Chase, 1993; Howard and Shaver, 1991). The herds averaged 12,995 kg of milk per cow with a range of 11,740 to 14,167 kg. Milk fat percent averaged 3.65% with a range of 3.3 to 3.95%. Milk protein percent averaged 3.15% with a range of 2.97 to 3.4%. One approach was to summarize the types of feeds used in these herds as follows:

Table 3.1 contains milk production and ration nutritional parameters for 8 high producing Holstein herds. As a comparison, NRC (1989) requirement guidelines are provided where possible. It must be emphasized that the data for the 8 herds in Table 3.1 are 'field' data rather than research data. For each of these herds, some pieces of information were estimated. However, this information does provide an overview of ration guidelines used in the field. The average daily total protein intake would be about 4.7 kg for the average DMI in Table 3.1. This would be adequate to support about 50 kg of milk production for second lactation dairy cows.

What is the basis for the high levels of milk production attained by these herds? The ration nutrient parameters in Table 3.1 are similar to those used in routinely formulating rations for high group rations. However, many of these herds are producing between 8,500 and 10,000 kg of milk per cow per year. The primary difference between these herds and those represented in Table 3.1 is DMI. These high producing herds have developed feeding management strategies to maximize intake. These observations are similar to those made in other high producing herds (Neilson *et al.*, 1983; Wood and Wilson, 1983).

Summary

The challenge of developing rations for high levels of milk production will persist in the future. As nutritionists, we should look forward to this opportunity. The key to attaining high levels of efficient and profitable milk production is to maximize DMI, optimize rumen fermentation and provide postruminal energy and protein supplements to meet tissue needs.

The integration of nutritional concepts into a balanced ration can most easily be done utilizing computer ration programs. There are a large number of programs being successfully utilized in the US. These programs are well suited to routine ration formulation. However, they are not designed to handle many of the newer nutritional concepts such as level of intake, rate of passage, digestion rates, and amino acids. Computer models have been developed which do consider the above concepts (Sniffen *et al.*, 1992). These models are in a continuing state of development, validation and field testing. It is anticipated that this type of approach will become more available in the future.

What guidelines do I use in formulating rations for high producing herds? The following guidelines are provided for consideration and discussion. They are based on both NRC (1989) and field experience. Some of these have not been fully defined by research data at this time.

1. Ration fibre levels
   a. ADF = 19–21% of DM
   b. NDF = 26–30% of DM
   c. Forage NDF = 20–22% of DM

2. NFC = 35–40% of DM

3. Fat = 5–7% of DM

4. Ration protein:
   a. Crude protein = 17–19% of DM
   b. Degradable protein = 60–65% of CP
   c. Undegradable protein = 35–40% of CP
   d. Soluble protein = 30–35% of CP

5. NE_l = 1.7–1.8 Mcal/kg of DM (7.1–7.5 MJ/kg of DM)

The milk produced using the above guidelines depends on DMI. As an example, assume that a ration with 18% CP is fed. Milk production would be 38 kg if DMI was 20 kg. However, if DMI is 25 kg, then milk production would be 49 kg on a protein basis for a mature cow. DMI needs to average

4% or more of bodyweight if high levels of milk production are to be attained. Individual cows will have DMIs in excess of 5% of bodyweight.

The principles outlined above are currently utilized in formulating rations for dairy herds in the US. These guidelines will continue to be modified as additional research information becomes available. A better understanding of nutrient digestion rates and amino acid nutrition are the key areas in which research is needed. Computer based ration formulation models will be needed to integrate and utilize these concepts. What are the limits to milk production in the dairy cow?

References

AFRC (1993) *Energy and Protein Requirements of Ruminants. An advisory manual prepared by the AFRC Technical Committee on Responses to Nutrients.* Wallingford, CAB International

Albright, J.L. (1992) Management and Behavior of the high producing dairy cow. *Proceedings of the Tri-State Dairy Nutrition Conference* p. 53

Chalupa, W. and Sniffen, C.J. (1994) Carbohydrate, protein and amino acid nutrition of lactating dairy cattle. In *Recent Advances in Animal Nutrition – 1994*, pp. 265–275. Edited by P.C. Garnsworthy and D.J.A. Cole. Nottingham: Nottingham University Press

Chase, L.E. (1991) Identifying optimal UIP sources to supply limiting amino acids. *Proceedings of the Cornell Nutrition Conference* p. 57

Chase, L.E. (1993) Developing nutrition programs for high producing dairy herds. *Journal of Dairy Science*, **76**, 3287

Clark, J.H., Klusmeyer, T.H. and Cameron, M.R. (1992) Microbial protein synthesis and flows of nitrogen fractions to the duodenum of dairy cows. *Journal of Dairy Science*, **75**, 2304

Colucci, P.E., Chase, L.E. and Van Soest, P.J. (1982) Level of feed intake and diet digestibility in dairy cattle. *Journal of Dairy Science*, **65**, 1445

Dhiman, T.R., Cadorniga, C and Satter, L.D. (1993) Protein and energy supplementation of high alfafa silage diets during early lactation. *Journal of Dairy Science*, **76**, 1945–1959

Erasmus, L.J., Prinsloo, J. and Meissner, H.H. (1988) The establishment of a protein degrability data-base for dairy cattle using the nylon bag technique I: Protein sources. *South African Journal of Animal Science*, **18**, 23–29

Erasmus, L.J., Botha, P.M., Cruywagen, C.W. and Meissner, H.H. (1994) Amino-acid profile and intestinal digestibility in dairy cows of rumen-undegradable protein from various feedstuffs. *Journal of Dairy Science,* **77,** 541–551

Faldet, M.A., Satter, L.D. and Broderick, G.A. (1992) Determining optimal heat treatment of soybeans by measuring available lysine chemically and biologically with rats to maximize protein utilization by ruminants. *Journal of Nutrition,* **122,** 151

Gunderson, S. (1992) How six top Wisconsin herds are fed. *Hoard's Dairyman,* **137,** 686

Holden, L.A., Glenn, B.P., Erdmann, R.A. and Potts, W.E. (1994) Effects of alfalfa and orchardgrass on digestion by dairy cows. *Journal of Dairy Science,* **77,** 2580–2594

Hoover, W.H. and Stokes, S.R. (1991) Balancing carbohydrates and proteins for optimum rumen microbial yield. *Journal of Dairy Science,* **74,** 3630

Howard, W.T. and Shaver, R.D. (1991) Rations fed on selected high Wisconsin herds – 1990. *Report by Department of Dairy Science,* Madison: University of Wisconsin

Jordan, E.R. and Fourdraine, R.H. (1993) Characterization of the management practices of the top milk producing herds in the country. *Journal of Dairy Science,* **76,** 3247

Kaiser, R.M. and Combs, D.K. (1989) Utilization of three maturities of alfalfa by dairy cows fed rations that contain similar concentrations of fiber. *Journal of Dairy Science,* **72,** 2301

Kawas, J.R., Jorgensen, N.A. and Danelon, J.L. (1991) Fibre requirements of dairy cows: optimum fibre level in lucerne-based diets for high producing cows. *Livestock Production Science,* **28,** 107

Kertz, A.F., Reutzel, L.F. and Thomson, G.M. (1991) DMI from parturition to midlactation. *Journal of Dairy Science,* **74,** 2290

Llamas-Lamas, G. and Combs, D.K. (1990) Effect of alfalfa maturity on fiber utilization by high producing dairy cows. *Journal of Dairy Science,* **73,** 1069

Llamas-Lamas, G. and Combs, D.K. (1991) Effect of forage to concentrate ratio and intake level on utilization of early vegetative alfalfa silage by dairy cows. *Journal of Dairy Science,* **74,** 526

Madsen, J. and Hvelplund, T. (1994) Prediction of in situ protein degradability in the rumen. Results of an European ringtest. *Livestock Production Science,* **39,** 201

Makoni, N.F., Shelford, J.A. and Fisher, L.J. (1994) Initial rates of degradation of protein fractions from fresh, wilted and ensiled alfalfa. *Journal of Dairy Science,* **77,** 1598–1603

McNiven, M.A., Hamilton, R.M.G., Robinson, P.H. and deLecuw, J.W. (1994) Effect of flame roasting on the nutritional quality of common cereal grains for non-ruminants and ruminants. *Animal Feed Science and Technology*, **47**, 31

Mertens, D.R. (1987) Predicting intake and digestibility using mathematical models of rumen function. *Journal of Animal Science*, **64**, 1548

National Research Council (NRC: 1989) *Nutrient Requirements of Dairy Cattle. 6th Revised Edition.* Washington: National Academy of Science

Neilson, D.R., Whittemore, C.T., Lewis, M., Alliston, J.C., Roberts, D.J., Hodgson-Jones, L.S., Mills, J., Parkinson, H. and Prescott, J.H.D. (1983) Production characteristics of high-yielding dairy cows. *Animal Production*, **36**, 321

Nocek, J.E. (1994) Effective management of rumen carbohydrates in rations for dairy cattle. *Proceedings 55th Minnesota Nutrition Conference*, p. 165.

Roseler, D.K., Fox, D.G., Chase, L.E. and Stone, W.C. (1993) Feed intake prediction and diagnosis in dairy cows. *Proceedings of the Cornell Nutrition Conference*, p. 216.

Satter, L.D., T.R. Dhiman and J.T. Hsu. (1994) Use of heat processed soybeans in dairy rations. *Proceedings of the Cornell Nutrition Conference*, p. 19.

Schwab, C.G. (1994) Amino acid requirements of lactating dairy cows. *Proceedings 55th Minnesota Nutrition Conference*, p. 179.

Sniffen, C.J., O'Connor, J.D., Van Soest, P.J., Fox, D.G. and Russell, J.B. (1992) A Net Carbohydrate and Protein System for evaluating cattle diets. II. Carbohydrate and protein availability. *Journal of Animal Science*, **70**, 3562

Stern, M.D., Calsmiglia, S. and Endres, M.I. (1994a) Dynamics of ruminal nitrogen metabolism and their impact on intestinal protein supply. *Proceedings of the Cornell Nutrition Conference*, p. 105

Stern, M.D., Varga, G.A., Clark, J.H., Firkins, J.L., Huber, J.T. and Palmquist, D.L. (1994b) Evaluation of chemical and physical properties of feeds that affect protein metabolism in the rumen. *Journal of Dairy Science*, **77**, 2762

Teller, E., Vanbelle, M., Foulon, M., Collignon, G. and Matatu, B. (1992) Nitrogen metabolism in the rumen and whole digestive tract of lactating dairy cows fed grass silage. *Journal of Dairy Science*, **77**, 1296–1304

Van Soest, P.J. and Fox, D.G. (1992) Discounts for net energy and protein – Fifth revision. *Proceedings of the Cornell Nutrition Conference*, p. 40

Waldo, D.R. (1986) Effect of forage quality on intake and forage-concentrate interactions. *Journal of Dairy Science*, **69**, 617

Weiss, W.P. and Shockey, W.L. (1991) Value of orchardgrass and alfalfa silages fed with varying amounts of concentrates to dairy cows. *Journal of Dairy Science*, **74**, 1933

Wood, P.D.P. and Wilson, P.N. (1983) Some attributes of very high-yielding British Friesian and Holstein dairy cows. *Animal Production*, **37**, 157

4

PREDICTION OF THE INTAKE OF GRASS SILAGE BY CATTLE

R.W.J. STEEN[1,2], F.J. GORDON[1,2], C.S. MAYNE[1,2], R.E. POOTS[3], D.J. KILPATRICK[2], E.F. UNSWORTH[2], R.J. BARNES[4], M.G. PORTER[1] AND C.J. PIPPARD[1]

[1] *Agricultural Research Institute of Northern Ireland, Hillsborough, Co. Down, UK*
[2] *Department of Agriculture for Northern Ireland, Newforge Lane, Belfast, UK*
[3] *Greenmount College of Agriculture and Horticulture, Antrim, UK*
[4] *Perstrop Analytical Ltd, Highfield House, Roxborough Way, Maidenhead, UK*

Introduction

Efficient production of high quality milk and beef is highly dependent on dairy cows and beef cattle receiving the correct intake of nutrients. However grass silage, which forms the basis of winter rations for the vast majority of dairy and beef cattle in the United Kingdom and other countries in North Western Europe, varies greatly in terms of chemical and biological composition due to the impact of factors such as sward type, fertilisation, climate and ensiling technique on the fermentation process in the silo and nutritive value. This in turn results in major variation in silage intake, which affects both the quantity and quality of milk and beef produced. The improvement in the genetic potential of dairy and beef cattle has increased the economic benefits of achieving optimal inputs of nutrients. There is also currently considerable concern about the impact of animal production on the environment, especially with regard to reducing losses of nitrogen and phosphate from animals, which adds further impetus to the need for accurate prediction of food and nutrient intakes. An effective method for predicting silage intake from its chemical and biological compositions is essential if animals are to be allocated the correct amounts of concentrates to provide the optimum inputs of nutrients.

Control of food intake in cattle

The commonly accepted theories of the control of food intake by cattle date back to at least 50 years ago when Lehmann (1941) proposed that the food intake of ruminants was limited by the amount of indigestible material that they were able to consume. Since then much research has contributed to the understanding of the factors controlling the food intake of cattle and the roles of both physiological and physical restriction of intake (e.g. Conrad, Pratt and Hibbs, 1964; Conrad, 1966; Dinius and Baumgardt, 1970). This led to the concept that the drive for cattle to consume food is related to their requirements for energy as determined by their genetic capacity for growth or milk production (e.g. Jones, 1972; Conrad *et al.*, 1964) and that restriction or control of intake of forage-based diets is determined largely by the rate of degradation of food in the rumen, which in turn limits the rate at which the animal can ingest additional food (Balch and Campling, 1962; Conrad, 1966). In addition to the rate of degradation in the rumen, food intake is also influenced by the extent to which an animal can moderate both its tolerance for different degrees of rumen fill and the rate of passage of digesta through the alimentary tract (Ketelaars and Tolkemp, 1992). The tolerance for rumen fill generally increases with increasing demand for nutrients while the rate of passage is generally increased by factors such as lactation and cold stress.

Despite the importance of the rate of degradation in the rumen, control of food intake by cattle is highly complex involving a wide range of factors in both the animal (such as milk yield, stage of lactation, pregnancy, previous plane of nutrition, body condition, age, breed and sex) and the feed (such as degradability, digestibility, rate of passage, physical form and chemical composition), and the totality of negative and positive effects and their interactions determine the overall level of food intake achieved by the animal (Ketelaars and Tolkamp, 1992). Furthermore patterns of food intake during pregnancy (especially late pregnancy) and lactation do not match requirements and there is evidence that this discrepancy is not caused purely by physical constraints, suggesting a more complex physiological regulation of food intake which may be related to the efficiency of utilization of metabolisable energy and changes in basal metabolism (Ketelaars and Tolkamp, 1992). From a recent review of the literature on factors affecting food intake and a re-assessment of data from earlier studies, Ketelaars and Tolkamp (1992) question the validity of the concept that cattle have a feeding drive which aims at maximising food intake to the extent that they can as far as possible achieve their genetic capacity for milk or meat production. The fact that cattle often exhibit compensatory growth and eat more, following a period of restricted growth, and that animals treated with a number of hormones have a higher level of performance and eat more, suggests that 'normal fed' or untreated animals do not achieve maximum intake or performance, and therefore, that if physical fill is a major factor controlling intake, its controlling effect can be considerably altered by the over-riding effect of the endocrine system of the animal. It is also not clear from the literature whether food intake increases in response to increased rate of passage or *vice-versa* (Journet and Remond, 1976). Intake has been shown to increase with increasing nitrogen content in the diet even when nitrogen content far exceeds that which might be expected to limit microbial fermentation in the rumen, and it has recently been suggested that feed characteristics which are commonly associated with the rumen filling effect of a feed also profoundly affect the metabolism of the animal which in turn may influence the hormonal control of food intake (Ketelaars and Tolkamp, 1992).

Gill, Rook and Thiago (1988) reviewed the results of their studies to examine the extent of rumen fill after individual meals in cattle offered grass and legume hays and grass silage once daily either *ad libitum* or in restricted quantities. The extent of rumen fill at the end of individual meals varied greatly between forages and between individual meals over the 24 hour period. From their results these authors concluded that both physical distension of the rumen and one or more end products of fermentation have important roles in the control of forage intake and that the relative import-ance of the different factors varies considerably between forages and over the 24 hour feeding period. Factors such as offering leguminous rather than grass hay reduced the extent of rumen fill at the end of an individual meal but this appeared to have little influence on overall intake, while offering silage rather than hay had a similar effect on meal size, but with the slower rate of disappearance of silage from the rumen, would appear to have had greater long-term effects on intake.

Control of silage intake

The intake of grass silage is generally lower than that of comparable hay or fresh forage (e.g. Demarquilly, 1973; Cushnahan and Gordon, 1995) although similar intakes of silage and comparable fresh forage have been recorded (Flynn, 1978; Cushnahan and Mayne, 1994). Silage intake has also been particularly variable and prediction of its intake highly problem-atic (Rook and Gill, 1990a). The lower intake of silage than that of fresh or dried forage has often been attributed to the end products of fermentation in silage. A number of authors have used simple or multiple regression analyses to examine the relationships between individual parameters or groups of parameters and voluntary intake of silage (e.g. Wilkins, Hutchin-son, Wilson and Harris, 1971; Wilkins, Fenlon, Cook and Wilson, 1978; Flynn, 1979; Jones, Larsen and Lanning, 1980; Lewis, 1981; Gill *et al.*,

1988; Rook and Gill, 1990; Rook, Dhanoa and Gill, 1990a). Wilkins *et al.*, (1971, 1978) analysed data for 70 and 142 silages respectively which were given to growing sheep over a wide range of experiments, while Jones *et al.*, (1980) analysed data for 11 silages given to non-pregnant, non-lactating ewes. Flynn (1979) analysed the data from 37 beef production experiments carried out in Ireland and Rook and Gill (1990) and Rook *et al.*, (1990a) used data from a wide range of experiments undertaken at three research centres in the UK in which silages were offered to growing beef cattle. Lewis (1981) reviewed the literature on dairy cow feeding experiments up to 1979, and analysed the data from 78 experiments while Gill *et al.*, (1988) analysed data from dairy cow feeding experiments collected over seven years at Hurley and which involved 206 lactations.

Factors affecting silage intake

EFFECT OF SILAGE DRY MATTER CONTENT

There have been significant and positive relationships between silage dry matter content and intake in studies which have involved a wide range of silage dry matter contents (Wilkins *et al.*, 1971; Flynn, 1979; Lewis, 1981; Gill *et al.*, 1988; Rook and Gill, 1990). The R^2 values for linear or quadratic relationships between dry matter content and intake ranged from 0.12 (Gill *et al*, 1988) to 0.44 (Rook and Gill, 1990). Rook and Gill (1990) reported a significant curvilinear relationship between dry matter content and intake which suggested that there was little increase in intake when dry matter content increased above 250 g/kg.

EFFECT OF TOTAL NITROGEN CONTENT

Wilkins *et al.*, (1971) and Gill *et al.*, (1988) obtained significant positive relationships between silage nitrogen content and intake, R^2 values for linear regressions being 0.33 and 0.13 respectively, while Wilkins *et al.*, (1978); Flynn (1979); Jones *et al.*, (1980) and Lewis (1981) were unable to detect a significant effect of total nitrogen content on silage intake. However, Rook and Gill (1990) found that the relationships between nitrogen content and silage intake were small in linear regression analyses but that nitrogen content had strong positive relationships with intake in all of their models when collinearity between nitrogen content and other parameters was removed.

AMMONIA-N CONTENT

Wilkins *et al.,* (1971, 1978), Flynn (1979), Lewis (1981), Gill *et al.,* (1988) and Rook and Gill (1990) all found significant negative relationships between ammonia-N as a proportion of total nitrogen in silage and silage intake. Wilkins *et al.,* (1971, 1978) reported that ammonia-N as a proportion of total-N accounted for 38 and 42% of the variation in the intake of non-formaldehyde-treated silages given to sheep, while Rook and Gill (1990) reported R^2 values of 0.16 to 0.44 for relationships between ammonia-N as a proportion of total N and intake by cattle in different data sets. The latter authors also found that the relationship between ammonia content and intake was curvilinear, with the effect on intake increasing as ammonia-N content increased. However, there is no evidence in the literature that ammonia content *per se* affects silage intake, and consequently the negative relationships between ammonia content and intake may be due to strong relationships between ammonia content and the causal agents. This is supported by the fact that the relationship between intake and ammonia content was not as strong when ammonia content was expressed as g/kg dry matter rather than on a g/kg total nitrogen basis. Furthermore, when Rook *et al.,* (1990a) carried out further analyses of the data, coefficients for ammonia-N became very small in all models when collinearity between ammonia content and other parameters was removed, indicating that there was little effect of ammonia content *per se* on intake. They also found a strong correlation between ammonia and volatile fatty acid contents, and the fact that the relationship between ammonia content and intake was removed when collinearity with volatile fatty acid content was removed led them to suggest that volatile fatty acid content was the casual agent.

PH AND TOTAL ACIDITY

Wilkins *et al.,* (1971, 1978), Flynn (1979), Jones *et al.,* (1980) and Lewis (1981) obtained no significant effect of pH on silage intake, while Gill *et al.,* (1988) and Rook and Gill (1990) obtained significant quadratic effects of pH, the general pattern being a positive relationship at low pH followed by a negative relationship as pH increased, with an inflexion at 4.15. However Rook and Gill (1990) state that this result should be treated with caution as there was confounding between the pH of the silages and level of concentrate supplementation. Any relevance of pH in the control of silage intake is also likely to vary according to the relative content of different acids and the buffering capacity of the silage. McLeod, Wilkins and Raymond (1970) examined the effects of adding sodium bicarbonate to silage and found that a reduction in acidity was accompanied by increased voluntary intake by sheep. However in recent studies involving a wide range of low dry matter silages ($< 220\,g$ dry matter/kg) with pH values in the range 3.5 to 4.0, offering sodium bicarbonate did not affect the intake of any of the silages when they were offered to cattle (Steen and Gordon, unpublished results; P. O'Keily, personal communication). Wilkins *et al.*, (1971, 1978) did not find any significant relationship between total acid content and intake by sheep while Rook and Gill (1990) obtained a significant negative relationship in only one out of three sets of data ($R^2 = 0.13$).

LACTIC ACID

The effects of lactic acid content of silage on intake have been variable. Wilkins *et al.*, (1971, 1978) did not obtain a significant relationship between lactic acid content and silage intake by sheep while Jones *et al.*, (1980) and Gill *et al.*, (1988) found significant negative correlations with the intakes of sheep and dairy cows ($R^2 = 0.07$) respectively. Rook and Gill (1990) obtained a significant positive relationship ($R^2 = 0.07$) in one of three sets of data. Furthermore when Rook and Gill (1990) further analysed their data and removed collinearity, the coefficients for lactic acid became very small, indicating that lactic acid content had little direct effect on intake.

VOLATILE FATTY ACIDS

In a number of studies there have been negative relationships between the contents of individual or total volatile fatty acids in silage and voluntary intake. Wilkins *et al.*, (1971) found that acetic acid content was negatively correlated with intake ($R^2 = 0.25$) while Gill *et al.*, (1988) obtained negative relationships between acetic and butyric acid contents and intake by dairy cows (R^2 values 0.15 and 0.03 respectively) and Rook and Gill (1990) obtained negative relationships between acetic, butyric and total volatile fatty acid content and intake (R^2 values 0.20, 0.31 and 0.24 respectively). When collinearity was removed from the data of Rook and Gill (1990), butyric acid content was still strongly correlated with intake, which would suggest that it had an important depressing effect on intake independent of other components within the silages.

There is some evidence that infusion into the rumen or addition to the diet of either acetic (Ulyalt, 1965; Simkins, Suttie and Baumgardt, 1965;

Buchanan-Smith, 1990) or butyric acid (Simkins *et al.*, 1965) can depress intake. However when combinations of two volatile acids or one acid plus another nutrient have been added, effects on intake have been inconsistent. For example, when butyric or lactic acids were added to the diet with acetic acid, the depressive effect of acetic added alone was eliminated or even reversed (Buchanan-Smith, 1990). Similarly, when propionic acid was infused into the rumen with acetic, the depression in intake due to acetic alone was eliminated (Bhattacharya and Warner, 1968). The effects of the concentrations of volatile acids in the diet on intake and the interactions between them and other dietary components are complex and the mechanisms involved are likely to include osmotic as well as chemoreceptors (Buchanan-Smith and Phillip, 1986).

AMINES AND OTHER NITROGENOUS COMPOUNDS

Few studies have examined the relationships between the contents of amines or other nitrogenous compounds such as gamma amino butyric acid in silages and their intake. Neumark, Bondi and Volcani (1964) found a negative correlation between tryptamine content of silage and intake, but it is not clear if this was due to the amine content *per se* or an association of higher amine content with other factors. In more recent studies the addition of high levels of putrescine, cadaverine or gamma amino butyric acid to the diet, or infused into the rumen did not significantly affect silage intake by cattle (Dawson and Mayne, 1994a, b). Similarly, Buchanan-Smith (1990) found that the addition of a mixture of putrescine, cadaverine and gamma amino butyric acid to the diet of sham-fed sheep did not depress intake. In fact, intermediate levels of addition tended to increase intake while the highest level of addition had little effect on intake, giving a quadratic response.

DIGESTIBILITY AND FIBRE CONTENT

Studies which have examined the relationship between the digestibility and intake of silage by cattle have generally shown significant positive relationships between digestibility and intake (e.g. Flynn, 1979; Lewis, 1981; Gill *et al.*, 1988; Rook and Gill, 1990). However in studies with sheep there has generally been no significant effect of digestibility on silage intake (e.g. Wilkins *et al.*, 1971, 1978; Jones *et al.*, 1980). This may indicate that compared to digestibility, other components such as products of fermentation and chop length are more important determinants of intake in sheep than in cattle.

Osbourn, Terry, Outen and Cammell (1974) found that intake was more closely related to neutral-detergent fibre (NDF) content than to digestible organic matter content of a range of forages. Mertens (1985) has suggested that NDF content should be a better parameter for predicting forage intake than either acid detergent fibre (ADF) or digestible organic matter content, as it is more closely related to cell wall content of the plant, rate of fibre digestion and hence to digesta volume. He subsequently suggested (Mertens, 1987) that data such as lignin concentration and particle size would also be needed to refine predictions in situations in which physical fill was limiting intake. Mertens' hypothesis is supported by the results of Jones *et al.* (1980) who found that the intakes of high dry matter silages were much more closely related to cell wall content and rate of dry matter disappearance in acid pepsin (correlation coefficients, − 0.50 and 0.54 respectively) than to ADF content or dry matter digestibility (correlation coefficients − 0.10 and 0.03 respectively).

EFFECTS OF CHOP LENGTH

Studies which have examined relationships between the composition of silages and intake have generally not included consideration of the effects of chop length on intake, possibly because there has been little objective data available on the chop length of the silages involved in the experiments. However there have been many experiments in which silages have been made with different types of equipment to produce silages with different nominal chop lengths and the effects of these on intake and animal performance have been recorded.

The results of an extensive series of studies at Hillsborough involving beef cattle, dairy cows and sheep have been reviewed by Steen (1984), Gordon (1986) and Chestnutt and Kilpatrick (1989) respectively. Grass was harvested with a flail forage harvester producing material with a mean chop length of approximately 85 mm and with a precision-chop harvester producing material with a mean chop length of 25 mm. Shorter chopping increased silage intake by 2%, 4% and 32% with dairy cows, beef cattle (400 to 550 kg live weight) and pregnant/lactating ewes respectively. In these studies the silages offered to the dairy cows, beef cattle and sheep were not harvested simultaneously from the same swards although they were harvested with the same equipment. However in one study, the results of which have been reported by Gordon (1982) and Apolant (1982), silages which had been harvested simultaneously from the same sward were offered to dairy cows and finishing lambs (35 kg live weight). In this case offering finely chopped, precision chopped silage rather than material which had been harvested with a double-chop machine or with a precision-chop machine with some of the blades removed to produce a longer particle length, did not affect the intake of dairy cows but increased the silage intake of lambs by 43 and 59% compared to the double chopped and long precision chopped silages respectively.

Prediction of silage intake

Wilkins *et al.* (1971, 1978), Flynn (1979), Jones *et al.* (1980), Lewis (1981), Rook and Gill (1990) and Rook *et al.* (1990a) have all produced multifactor relationships to predict the silage dry matter intake of sheep or cattle. However each of these models has been based on historical data from a range of experiments and consequently the effects of factors within the silages on intake have been confounded with other factors such as the breed, age, weight, physiological state, previous nutritional history and body condition of the animals, the prevailing environmental conditions at the time of each experiment, feeding management and in some cases also with the effects of level and type of concentrate supplementation and milk yield and stage of lactation of the animals. These confounding effects have reduced the accuracy of the predictive relationships. The value of the relationships produced in a number of the studies has also been limited by a lack of comprehensive data on the chemical compositions of the silages, a limitation which was recognised by Lewis (1981) and Rook, Dhanoa and Gill (1990b). In most cases there has also been no attempt made to validate the models in terms of their ability to predict silage intake.

More recently, work undertaken by Offer, Dewhurst and Thomas (1994), which involved the determination of the intakes of 57 silages by sheep, resulted in stronger relationships between the chemical and biological compositions of the silages and intake. Silage intake by lambs was best predicted by an eight-term principle components regression involving the live weight of the animals and digestible organic matter, crude protein, soluble protein, lactic acid and total volatile acid contents of the silages, and total volatile acids as a proportion of total acids and soluble protein as a proportion of crude protein in the silages. This relationship had an R^2 value of 0.81. However when 16 of the silages were fed to dairy cows the model for lambs poorly predicted silage intake by the cows ($R^2 = 0.31$). When additional parameters were taken into consideration the R^2 value for prediction of intakes by cows from intakes by lambs was improved to 0.79. Offer *et al.* (1994) did not present data on the validation of this model and no information was given on the chop length of the silages or on whether or not the data set involved silages of more than one chop length. This is particularly important in view of the very different effects of chop length on the intakes of silages by sheep and cattle as discussed earlier. There is also further evidence that data relating to the intake of silages by sheep is unlikely to be applicable to cattle (Cushnahan, Gordon, Ferris, Chestnutt and Mayne, 1994).

Direct prediction of intake of forages using near infrared reflectance spectroscopy

Near infrared reflectance spectroscopy (NIRS) has been used successfully for several years to predict, not only the chemical composition of feedstuffs, but also biological parameters such as organic matter digestibility (Norris, Barnes, Moore and Stenk, 1976; Barber, Givens, Kridis, Offer and Murray, 1990). Although there would appear to be no information in the literature on the direct prediction of silage intake using NIRS, it has been used to predict the intake of dried forages (Coelho, Hembry, Barton and Saxton, 1988). Coelho *et al.* (1988) reported an R^2 value of calibration of 0.84 for a multiple-linear regression relationship between second derivative NIRS data and the intake of hays by steers. The R^2 values for relationships between NIRS data and crude protein, NDF and ADF contents and dry matter digestibility of the hays used in the same experiment were 0.97, 0.98, 0.96 and 0.69 respectively. The lower R^2 values for the *in vivo* determinations reflect the greater variability in animal responses to the forages (Coelho *et al.*, 1988). The use of NIRS to directly predict intake also has the advantage that the problems of major collinearity between factors should be minimised by the use of partial least squares analysis or differencing between wavelengths.

Research programme at Hillsborough on the prediction of silage intake

In view of the limitations of previous models based on historical data or data obtained with sheep to predict silage intake by dairy cows and beef cattle, a research programme was initiated at Hillsborough in 1992 with the aim of improving the prediction of silage intake by cattle. The intakes of 136

silages from Northern Ireland farms were determined using 192 individu-
ally-fed beef cattle with a mean live weight of 415 kg. The silages were
selected on the basis of their pH and dry matter, ammonia and metabolisable energy (predicted by NIRS) contents with the aim of obtaining a
wide range of chemical compositions. Approximately seven tonnes of each
silage was brought to the Institute by covered lorry. Each lot of silage was
mixed in a mixer wagon to achieve uniformity and was then stored in
polythene-lined, sealed and evacuated boxes as described by Pippard,
Porter, Steen, Gordon, Mayne, Agnew, Unsworth and Kilpatrick (1995)
until feeding one to four weeks later. There was no deterioration of the
silages during storage and their chemical composition remained constant
(Pippard *et al.*, 1995). They were offered to the cattle in two linked changeover design experiments, with each silage being offered as the sole feed to
ten animals for a period of two weeks. Eight silages were offered in each of
17 periods and in addition a further 16 animals were offered a standard hay
in each period to enable variation in intake due to periods to be removed.
Detailed chemical and biological compositions of the silages were determined including *in vivo* digestibility and the use of NIRS and electrometric
titration. The ranges in the chemical compositions and intakes of the 136
silages are summarised in Table 4.1. These confirm that the silages had a
wide range of chemical compositions.

Thirteen of the 136 silages were also offered unsupplemented to dairy
cows to provide a basis by which the intake system developed from the data
produced with beef cattle could be translated for use with dairy cows. In
addition sixteen of the silages were offered to dairy cows and growing beef
cattle with a range of concentrates. Each silage was supplemented with
high-starch and high-fibre concentrates, each with three protein contents
and at three levels of supplementation, to examine the interactions between
silage type and concentrate type/level of supplementation in terms of their
effects on silage intake.

Table 4.1 RANGES OF CHEMICAL COMPOSITIONS AND INTAKES OF SILAGES

Parameter	Range	SD
Dry matter (g/kg)	155 to 413	43.2
Crude protein (g/kg DM)	77 to 212	24.5
Ammonia-N (g/kg total N)	45 to 385	63.5
pH	3.5 to 5.5	0.40
Metabolisable energy (MJ/kg (predicted by NIRS)	8.8 to 12.3	0.8
Silage dry matter intake		
(kg/day)	4.3 to 10.9	1.13
(g/kg $w^{0.75}$)	45 to 113	12.0

Data from the main study involving the 136 silages offered without supplementation have been used to develop relationships between individual parameters or groups of parameters and intake using simple and multiple-regression analyses. Samples of each of the silages which had been dried at 85°C for 20 hours and allowed to equilibrate under normal laboratory conditions were also presented to a Perstrop Analytical Near Infrared Spectrometer. The spectrum for each sample was recorded over the full range of 400 to 2500 nm at 2 nm intervals. In addition the samples were scanned immediately following re-drying. Modified partial least squares regression analyses was used to investigate the relationship between the spectra for each sample and silage intake. Standard normal variate and detrend mathematical transforms were used to minimise any effects of particle size, temperature or residual moisture. Calibration models were

Table 4.2   R² VALUES FOR RELATIONSHIPS BETWEEN INDIVIDUAL PARAMETERS AND INTAKE

Parameter (g/kg dry matter unless otherwise stated)	R^2 linear	quadratic
Dry matter	0.21	0.22
Crude protein	0.19	0.32
Acid insoluble N	0.08	0.19
Soluble-N (total N minus acid insoluble N)	0.19	0.28
Amino acid-N	0.15	0.16
True protein	0.03	0.09
Ammonia-N (g/kg total N)	0.10	0.09
Soluble-N minus ammonia-N	0.34	0.35
Total N minus ammonia-N	0.33	0.35
Lactic acid	0.01	0.09
Acetic acid	0.05	0.08
Propionic acid	0.07	0.08
Butyric acid	0.04	0.03
Total volatile fatty acids	0.07	0.07
Lactic acid as a proportion of total acids	0.08	0.08
Ethanol	0	0.06
Propanol	0.08	0.08
pH	0	0.08
Buffering capacity (m equiv/kg DM)	0.02	0.02
Total free acids	0.02	0.03
Ether extract	0.12	0.15
Residual sugars	0.08	0.22
Neutral detergent fibre	0.38	0.38
Acid detergent fibre	0.36	0.35
Acid detergent lignin	0.34	0.49
Ash	0	0.08
Digestible organic matter content (predicted by NIRS)	0.19	0.19
Organic matter digestibility (*in vivo*)	0.30	0.29
Dry matter degradability	0.28	0.29
Electrometric titration	0.53	

developed separately for the equilibrated and re-dried sample sets and also for the combined sample set.

The relationships between individual parameters and intake have been produced to provide an overview of the extent to which individual parameters were correlated with intake. The R^2 values for these relationships are presented in Table 4.2. However it should be noted that, while these results provide an overall picture of the main groups of parameters which are closely related to intake and those which are poorly correlated with intake, excessive emphasises should not be placed on the significance of individual values due to the likelihood of the existence of collinearity between individual parameters.

DRY MATTER CONTENT

The increase in dry matter intake as dry matter content increased is in agreement with the results of previous studies (Wilkins *et al.*, 1971; Flynn, 1979; Lewis, 1981; Gill *et al.*, 1988; Rook and Gill, 1990). However there was no indication of major curvilinearity in the response, in contrast to the findings of Rook and Gill (1990) who found that there was little further increase in intake when dry matter content increased above 250 g/kg.

NITROGEN COMPONENTS

The relationship between crude protein content and intake was much stronger than that found in most previous studies, but was in agreement with the results of Wilkins *et al.* (1971). The pronounced curvilinear nature of the relationship would indicate that there was a strong positive relationship with intake at low nitrogen contents, little increase in intake at nitrogen contents above 140 g/kg dry matter, and a slight decrease above 160 g/kg. The response in intake to increasing protein content is likely to have been at least partly due to low nitrogen supply from the low protein silages reducing digestibility of the fibre components of the diet and hence rate of disappearance and intake due to rumen fill effects. This is in line with the strong relationships between intake and soluble-N, soluble-N minus ammonia-N and amino acid nitrogen contents. Alternatively, an inadequate absorption of nitrogen from the small intestine and supply of nitrogen to the animal's tissues may create an effective energy surplus and reduce intake (Egan, 1965). This is a possible contributing factor to the positive relationship between insoluble-N content and intake. The relationship between ammonia content and intake was weaker than those reported previously (Wilkins *et al.*, 1971, 1978; Rook and Gill, 1990). Furthermore the fact that the R^2 value for the relationship between ammonia-N content, expressed on a g/kg total N basis and intake was 0.10, while the R^2 value was only 0.01 when ammonia content was expressed on a g/kg dry matter basis, would indicate that the former relationship was not due to ammonia content *per se*, but rather to an association between ammonia content and some other parameter(s) within the silages which was the causal agent.

DIGESTIBILITY, DEGRADABILITY AND FIBRE FRACTIONS

The positive relationship between *in vivo* organic matter digestibility and intake is in agreement with the results of previous studies involving cattle (Flynn, 1979; Lewis, 1981; Gill *et al.*, 1988; Rook and Gill, 1990). However the relationship was much stronger than those reported previously. For example, Gill *et al.* (1988) and Rook and Gill (1990) reported R^2 values of 0.03 to 0.12 for relationships between digestibility and intake compared to the value of 0.30 in the Hillsborough studies. The relationship between digestibility and intake indicates that intake increased by 1.5% of the mean intake of all 136 silages for each 10 g/kg increase in digestibility. This is close to the response in intake of 1.7% per 10 g/kg increase in digestibility reported by Steen (1988) from a review of the results of eight experiments which examined the effects of predetermined differences in digestibility (by cutting grass at different stages of growth), on the intake and performance of beef cattle offered the silages as the sole diet. The fact that very similar relationships between digestibility and intake have been obtained for experimentally produced silages and silages from farms is particularly reassuring, and provides strong evidence for the validity of this relationship.

Osbourn *et al.* (1974) found that intake was more highly correlated with NDF content than with digestible organic matter content, while Mertens (1985) suggested that NDF content would be a better predictor of intake than either ADF or digestible organic matter content, as NDF is more closely related to the cell wall content of the plant and hence to rate of digestion and digesta volume. However in the Hillsborough study relationships between intake and ADF and NDF were similar ($R^2 = 0.36$ and 0.38) while the quadratic relationship for acid detergent lignin (ADL) was strongest ($R^2 = 0.49$). Intake was also strongly related to *in vivo* dry matter degradability, R^2 values being 0.28 and 0.26 for assumed rumen outflow rates of 0.05 and 0.08 per hour respectively. Data on degradabilities of NDF and ADF and hemicellulose in this study are not yet available but should be of particular interest in terms of contributing to the understanding of the parameters which controlled intake in this study.

LACTIC AND VOLATILE FATTY ACID CONTENTS

The contents of lactic, acetic, propionic and butyric acids and total volatile fatty acids were very weakly related to intake ($R^2 = 0.01$ to 0.09). This was despite the fact that there were wide ranges in the contents of these acids in the silages, these being 0 to 144 (S.D. 34.7); 4 to 63 (S.D. 12.7); 0 to 13 (S.D. 3.1) and 0 to 32 (S.D. 7.0) g/kg dry matter respectively. As discussed previously, the mechanisms through which an increase in the concentrations of lactic or volatile fatty acids in the diet may affect intake are complex and poorly understood. However, unless there was strong collinearity between these and other factors within the silages, the results of the studies at Hillsborough would indicate that the concentrations of lactic and volatile fatty acids in silage or their relative proportions, are of little importance in determining intake.

PH AND TOTAL ACIDITY

The pH, total free acid content and buffering capacity of the silages were also very weakly related to intake (R^2 0 to 0.08). The quadratic relationship between pH and intake, followed a similar trend to that produced by Rook and Gill (1990), in that there was a slight positive relationship at low pH followed by a negative relationship at high pH and maximum intake at pH 4.3. However the relationship was much weaker than that reported by Rook and Gill (1990).

OTHER PARAMETERS

Relationships between ethanol and propanol contents and intake were weak ($R^2 = 0.06$ and 0.08 respectively). The relationship with residual sugar content was strongly curvilinear ($R^2 = 0.22$), with a strong positive relationship at low sugar contents and a decreasing response in intake at higher sugar contents. However it is not clear to what extent relationships with parameters such as oil and sugar content are direct effects and to what extent they are due to collinearity with other factors.

The best relationship between electrometric titration and intake had an

R^2 value of 0.53. This would appear to have been largely attributable to strong relationships between the contents of total soluble-N, amino acid-N and reducing sugars on a fresh basis (estimated by electrometric titration) and intake. The R^2 values for the relationships between these three parameters and intake were 0.35, 0.32 and 0.30 respectively. These were partly attributable to the relationship between silage dry matter content and intake, as the relationships for the contents of soluble-N, amino acid-N and reducing sugars expressed on a dry matter basis had R^2 values of only 0.17, 0.16 and 0.04 respectively.

MULTIPLE REGRESSION RELATIONSHIPS

Multiple regression analyses have been used to produce relationships between intake and a range of chemical parameters in the silages. For this purpose the 136 silages were divided into two groups, one of 91 and the other of 45 silages. Relationships were produced using the 91 silages and the other 45 silages were used for validation of the relationships. The best of these relationships had an R^2 value of calibration of 0.69 and an R^2 of prediction of 0.63. However this relationship involved parameters such as ether extract and ADL which would not normally be determined in routine silage analyses. A relationship involving only dry matter, nitrogen and ammonia contents and pH had an R^2 of calibration of 0.63 and an R^2 of prediction of 0.56. When ADF content and electrometric titration were added to these four parameters, the best relationship had an R^2 of calibration of 0.65 and an R^2 of prediction of 0.61.

DIRECT PREDICTION OF INTAKE BY NIRS

The best overall relationship was between NIR spectra and intake. When calibration models were developed separately for the equilibrated and re-dried sample sets and for the combined sample set, R^2 values were 0.85, 0.85 and 0.86 respectively and the standard error of calibration was 4.3 g/kg $W^{0.75}$ for all three relationships. Two approaches to validation of the relationships were undertaken. Firstly cross validation of the model for the re-dried samples was also undertaken. For this purpose the spectra for one of the 136 silages was removed from the set and a model produced using the spectra for the remaining 135 silages. The intake of the remaining silage was then predicted using the model based on the other 135 samples. This procedure was repeated for each of the 136 silages. The R^2 of prediction was 0.73 and the standard error of cross validation was 6.1 g/kg $W^{0.75}$. In the second approach to validation the spectra for the equilibrated and re-dried samples were divided at random into two batches, one of 94 and one of 42. A calibration model was produced using the larger set of spectra and this was then used to predict the intakes of the other 42 silages. This procedure gave an R^2 of prediction of 0.71 and a standard error of prediction of 5.5 g/kg $W^{0.75}$ Although this standard error, at 7.7% of the mean intake of the 136 silages, is numerically large in comparison with the standard errors obtained when the chemical composition or organic matter digestibility of silages have been predicted by NIRS, it is not large relative to the accuracy with which silage intake can be determined *in vivo*. For example, the typical standard error for the determination of organic matter digestibility is 0.007 while Barber *et al.* (1990) and Baker, Givens and Deaville (1994) reported standard errors of prediction of 0.026 and 0.0235 respectively for the prediction of organic matter digestibility by NIRS. Thus the standard error of prediction of organic matter digestibility by NIRS is 3.4 times the typical standard error of determination. By comparison if the typical standard error for the determination of silage dry matter intake is taken as 2.8 g/kg $W^{0.75}$ (i.e. 4% of the mean intake), then the standard error of prediction of intake by NIRS was less than twice the standard error of determination.

APPLICATION OF THE PREDICTION OF SILAGE INTAKE BY NIRS

Prediction of silage intake by cattle has now been commissioned on a commercial basis in Northern Ireland, and routine laboratory analyses of silage samples now includes a prediction of the intake potential of each silage. Intake is predicted on a g dry matter/kg metabolic live weight basis. This value is then used directly to calculate the potential intake by beef cattle of various live weights. In the case of dairy cows, a conversion relationship has been produced, based on the intakes of the 13 silages which were offered unsupplemented to both beef cattle and dairy cows, which enables the intake value to be converted into an actual predicted daily intake for dairy cows. Feeding models are currently being produced for both beef cattle and dairy cows which will use the predicted silage intakes to calculate total metabolisable energy intakes and predict performance for different inputs of concentrates, or alternatively, predict the inputs of concentrates required to sustain given levels of performance.

INTERACTION BETWEEN CONCENTRATE AND SILAGE TYPE

As outlined previously 16 of the silages used in the main study were also offered to dairy cows and beef cattle with a wide range of levels and types of concentrate supplementation. Statistical analyses of the data for these 16 silages which were supplemented with different levels and types of concentrates have shown linear relationships between concentrate intake level and silage intake. Overall there was a tendency for substitution rate to be higher at the higher inputs of concentrates but this curvilinear effect was not significant. The mean substitution rate decreased with increasing crude protein content in the concentrates, the mean overall substitution rates being 0.46, 0.43 and 0.36 for the concentrates containing 120, 190 and 260 g crude protein/kg respectively. There was no overall difference in the mean substitution rates for the high-starch and high-fibre concentrates. However there were important interactions between silage type and concentrate type in terms of substitution rate. For example, silages A and B had very different patterns of fermentation but similar intakes when offered without supplementation as shown in Table 4.3. Yet silage A had a much lower intake (higher substitution rate) when supplemented with the high-starch concentrate than when supplemented with the high-fibre concentrate, while for silage B the two concentrate types had the opposite effect on silage intake. Prediction of these interactions is vitally important and research in this area is currently being pursued further at Hillsborough. In addition research is also aimed at elucidating the mechanisms involved in producing these interactions, with a view to providing a scientific basis on which different types of silages can be supplemented with the most appropriate type of concentrates.

CONCLUSIONS

Silage intake was strongly related to the contents of a number of nitrogen and fibre components, digestibility, dry matter degradability and residual water soluble carbohydrate content ($R^2 = 0.19$ to 0.49).

Table 4.3   INTERACTION BETWEEN SILAGE TYPE AND CONCENTRATE TYPE

	Silage	
	A	*B*
Silage intake without supplement (kg DM/day)	11.2	11.4
Substitution rate		
High-starch concentrate	0.48	0.39
High-fibre concentrate	0.28	0.55

Intake was either unrelated to, or only weakly related to, pH, buffering capacity, total acidity and the concentrations of lactic acid and volatile fatty acids ($R^2 < 0.09$).

The use of NIRS is a very simple and low-cost method of providing the most accurate prediction of silage intake.

Current research is elucidating the interactions between silage type and concentrate type, in terms of substitution rate, and this must be central to any approach to designing diets, particularly for dairy cows receiving moderately high levels of concentrates.

Acknowledgements

The authors wish to express their thanks to the Northern Ireland Grain Trade Association, the Milk Marketing Board for Northern Ireland and Strathroy Milk Marketing Ltd for financial support for the programme of research at Hillsborough on the prediction of silage intake.

References

Apolant, S.M. (1982) *A study on the value of grass silage-based diets for sheep with particular reference to the breeding ewe.* PhD Thesis, The Queen's University of Belfast

Baker, C.W., Givens, D.I. and Deaville, E.R. (1994) Prediction of organic matter digestibility *in vivo* of grass silage by near infrared reflectance spectroscopy: effect of calibration method, residual moisture and particle size. *Animal Feed Science and Technology*, **50**, 17–26

Balch, C.C. and Campling, R.C. (1962) Regulation of food intake by ruminants. *Nutrition Abstracts and Reviews*, **32**, 669–686

Barber, G.D., Givens, D.I., Kridis, M.S., Offer, N.W. and Murray, I. (1990) Prediction of the organic matter digestibility of grass silage. *Animal Feed Science and Technology*, **28**, 115–128

Bhattacharya, A.N. and Warner, R.G. (1968) Effect of propionate and citrate on depressed feed intake after intraruminal infusions of acetate in dairy cattle. *Journal of Dairy Science*, **51**, 1091–1094

Buchanan-Smith, J.G. (1990) An investigation into palatability as a factor responsible for reduced intake of silage by sheep. *Animal Production*, **50**, 253–260

Buchanan-Smith, J.G. and Phillip, L.E. (1986) Food intake in sheep following intraruminal infusion of extracts from lucerne silage with particular reference to organic acids and products of protein degradation. *Journal of Agricultural Science, Cambridge,* **106**, 611–617

Chestnutt, D.M.B. and Kilpatrick, D.J. (1989) Effect of silage type and concentrate supplementation on the intake and performance of breeding ewes. *62nd Annual Report, Agricultural Research Institute of Northern Ireland,* pp. 21–30

Coelho, M., Hembry, F.G., Barton, F.E. and Saxton, A.M. (1988) A comparison of microbial, enzymatic, chemical and near-infrared reflectance spectroscopy methods in forage evaluation. *Animal Feed Science and Technology,* **20**, 219–231

Conrad, H.R. (1966) Symposium on factors influencing the voluntary intake of herbage by ruminants: physiological and physical factors limiting feed intake. *Journal of Animal Science,* **25**, 227–235

Conrad, H.R., Pratt, A.D. and Hibbs, J.W. (1964) Regulation of feed intake in dairy cows. I. Change in importance of physical and physiological factors with increasing digestibility. *Journal of Dairy Science,* **47**, 54–62

Cushnahan, A. and Gordon, F.J. (1995) The effects of grass preservation on intake, digestibility and rumen degradation characteristics. *Animal Science* (In press)

Cushnahan, A., Gordon, F.J., Ferris, C.P.W., Chestnutt, D.M.B. and Mayne, C.S. (1994) The use of sheep as a model to predict the relative intakes of silages by dairy cattle. *Animal Production,* **59**, 415–420

Cushnahan, A. and Mayne, C.S. (1994) Effects of ensilage and silage fermentation pattern on the intake and performance of lactating dairy cows. *Animal Production,* **58**, 427 (Abstract)

Dawson, L.E.R. and Mayne, C.S. (1994a) The effects of either dietary addition or intraruminal infusion of amines or juice extracted from grass silage on the voluntary intake of steers offered grass silage. *Animal Production,* **58**, 427 (Abstract)

Dawson, L.E.R. and Mayne, C.S. (1994b) The effect of infusion of amines and gamma amino butyric acid on the intake by steers of grass silage differing in lactic acid content. *Proceedings of the 4th Research Meeting, British Grassland Society,* pp. 69–70

Demarquilly, C. (1973) Chemical composition, fermentation characteristics, digestibility and voluntary intake of forage silages: changes compared to the initial green forage. *Annales de Zootechnie,* **22**, 199–218

Dinius, D.A. and Baumgardt, B.R. (1970) Regulation of food intake in ruminants. 6. Influence of caloric density of pelleted rations. *Journal of Dairy Science,* **53**, 311–316

Egan, A.R. (1965) Nutritional status and intake regulation in sheep. III. The relationship between improvement in nitrogen status and increase in voluntary intake of low-protein roughages by sheep. *Australian Journal of Agricultural Research*, **16**, 463–472

Flynn, A.V. (1978) The effect of ensiling on the beef production potential of grass. *Proceedings of the 5th Silage Conference, Ayr*, pp. 48–49

Flynn, A.V. (1979) The effect of silage dry-matter digestibility *in vitro* on live-weight gain and carcass gain by beef cattle fed silage *ad libitum*. *Animal Production*, **28**, 423 (Abstract)

Gill, M., Rook, A.J. and Thiago, L.R.S. (1988) Factors affecting the voluntary intake of roughages by the dairy cow. In *Nutrition and Lactation in the dairy cow*, pp. 262–279. Edited by P.C. Garnsworthy. London: Butterworths

Gordon, F.J. (1982) The effects of degree of chopping grass for silage and method of concentrate allocation on the performance of dairy cows. *Grass and Forage Science*, **37**, 59–65

Gordon, F.J. (1986) The influence of system of harvesting grass for silage on milk production. *Jubilee Report, Agricultural Research Institute of Northern Ireland*, pp. 13–22

Jones, G.M. (1972) Chemical factors and their relation to feed intake regulation in ruminants: a review. *Canadian Journal of Animal Science*, **52**, 207–239

Jones, G.M., Larsen, R.E. and Lanning, N.M. (1980) Prediction of silage digestibility and intake by chemical analyses or *in vitro* fermentation techniques. *Journal of Dairy Science*, **63**, 579–586

Journet, M. and Remond, B. (1976) Physiological factors affecting the voluntary intake of feed by cows: a review. *Livestock Production Science*, **3**, 129–146

Ketelaars, J.J.M.H. and Tolkemp, B.J. (1992) Toward a new theory of feed intake regulation in ruminants. 1. Causes of differences in voluntary feed intake: critique of current views. *Livestock Production Science*, **30**, 269–296

Lehmann, F. (1941) Die Lehre vom Ballast. *Jeitschrift fur Tierernahrung und Futtermittelkunde*, **5**, 155–173

Lewis, M. (1981) Equations for predicting silage intake by beef and dairy cattle. *Proceedings of the Sixth Silage Conference, Edinburgh*, pp. 35–36

McLeod, D.S., Wilkins, R.J. and Raymond, W.F. (1970) The voluntary intake by sheep and cattle of silages differing in free-acid content. *Journal of Agricultural Science, Cambridge*, **75**, 311–319

Mertens, D.R. (1985) Effect of fibre on feed quality for dairy cows. *Proceedings of the 46th Minnesota Nutrition Conference*, pp. 209–224

Mertens, D.R. (1987) Predicting intake and digestibility using mathematical models of ruminal function. *Journal of Animal Science*, **64**, 1548–1558

Neumark, H., Bondi, A. and Volcani, R. (1964) Amines, aldehydes and keto-acids in silages and their effect on food intake in ruminants. *Journal of the Science of Food and Agriculture*, **15**, 487–492

Norris, K.H., Barnes, R.F., Moore, J.E. and Stenk, J.S. (1976) Predicting forage quality by infrared reflectance spectroscopy. *Journal of Animal Science*, **43**, 889–897

Offer, N.W., Dewhurst, R.J. and Thomas, C. (1994) The use of electro-metric titration to improve the routine prediction of silage intake by lambs and dairy cows. *Animal Production*, **58**, 427 (Abstract)

Osbourn, D.F., Terry, R.A., Outen, G.E. and Cammell, S.B. (1974) The significance of a determination of cell walls as the rational basis for the nutritive evaluation of forages. *Proceedings of the 12th International Grassland Congress, Moscow*, **Vol III**, pp. 374–380

Pippard, C.J., Porter, M.G., Steen, R.W.J., Gordon, F.J., Mayne, C.S., Agnew, R.E., Unsworth, E.F. and Kilpatrick, D.J. (1995) A method for obtaining and storing uniform silage for feeding experiments. *Animal Feed Science and Technology*, (Submitted)

Rook, A.J. and Gill, M. (1990) Prediction of the voluntary intake of grass silages by beef cattle. 1. Linear regression analyses. *Animal Production*, **50**, 425–438

Rook, A.J., Dhanoa, M.S. and Gill, M. (1990a) Prediction of the voluntary intake of grass silages by beef cattle. 2. Principal component and ridge regression analyses. *Animal Production*, **50**, 439–454

Rook, A.J., Dhanoa, M.S. and Gill, M. (1990b) Prediction of the voluntary intake of grass silages by beef cattle. 3. Precision of alternative prediction models. *Animal Production*, **50**, 455–466

Simkins, K.L., Suttie, J.W. and Baumgardt, B.R. (1965) Regulation of food intake in ruminants. 4. Effects of acetate, propionate, butyrate and glucose on voluntary food intake in dairy cattle. *Journal of Dairy Science*, **48**, 1635–1642

Steen, R.W.J. (1984) The effects of wilting and mechanical treatment of grass prior to ensiling on the performance of beef cattle and beef output per hectare. *57th Annual Report, Agricultural Research Institute of Northern Ireland*, pp. 21–32

Steen, R.W.J. (1988) Factors affecting the utilization of grass silage for beef production. In. Efficient Beef Production from Grass. (Ed. J. Frame). *Occasional Publication No. 22, British Grassland Society*, pp. 129–139

Ulyatt, M.J. (1965) The effects of intraruminal infusions of volatile fatty acids on food intake of sheep. *New Zealand Journal of Agricultural Science*, **8**, 397–408

Wilkins, R.J., Fenlon, J.S., Cook, J.E. and Wilson, R.F. (1978) A further analysis of relationships between silage composition and voluntary intake by sheep. *Proceedings of the 5th Silage Conference, Ayr*, pp. 34–35

Wilkins, R.J., Hutchinson, K.J., Wilson, R.F. and Harris, C.E. (1971) The voluntary intake of silage by sheep. 1. Interrelationships between silage composition and intake. *Journal of Agricultural Science, Cambridge*, **77**, 531–537

II

Manufacturing and Legislation

5

EFFECT OF SOME MANUFACTURING TECHNOLOGIES ON CHEMICAL, PHYSICAL AND NUTRITIONAL PROPERTIES OF FEED

A.G.J. VORAGEN[1], H. GRUPPEN[1], G.J.P. MARSMAN[1] AND A.J. MUL[2]

[1] *Department of Food Science, Wageningen Agricultural University, P.O. BOX 8129, 6700 EV, Wageningen, The Netherlands*
[2] *Nutreco Agri Specialties, Veerstraat 38, P.O. BOX 180, 5830 AD Boxmeer, The Netherlands*

Introduction

Raw materials for the production of feed are complex mixtures of innumerable chemical compounds which can undergo many chemical, biochemical, microbial and physical changes as a result of processing and storage. These changes influence the digestibility and nutritional quality of the feed. Apart from chemical composition, concentration of reactants and physical structures, the extent of these changes also depend on intrinsic factors like pH, water activity or presence of oxygen and process parameters like temperature, residence time, shear forces, etc.

These parameters are even more important in the production of food, which has to meet higher quality standards regarding microbiological safety, absence of toxic compounds, nutritional value and sensoric properties than feed. Most of the changes which can take place in food manufacturing, favourable or unfavourable for the overall quality of the food, are known, mostly qualitativly but increasingly also on a quantitative basis. Although raw materials and composition of feed, manufacturing technologies, microbial, toxicological and nutritional standards differ greatly from food, the principles for chemical, biochemical and physical changes are basically the same. Therefore, when food scientists look at feed it is obvious that they will try to extrapolate their knowledge of reactions taking place in food to feed systems. This paper concentrates on chemical, physical and nutritional changes due to thermo-mechanical technologies, e.g. steam conditioning, pelleting, expansion, extrusion, flaking, and cooling and the impact they have on properties of feedstuffs like pellet quality, flavour, taste, digestibility, nutritional value, adverse physiological reactions and on shelf-life.

COMPONENTS OF FEED RAW MATERIALS

The major constituents of most feedstuffs are carbohydrates, lipids and proteinaceous material including amino acids and peptides. As the different members of these groups of constituents differ in their chemical reactivity during processing they have been subdivided into different groups, as presented in Table 5.1. Also relevant in this context are minor constituents like bio-active compounds (enzymes, lectins, protease inhibitors, antigens) and other antinutritional factors like phenolics, glycosinolates, phytates, etc.

Carbohydrates

Carbohydrates can be classified according to their molecular size into low molecular weight mono-, and oligosaccharides and as high molecular

Table 5.1.   COMPONENTS OF FEED AND FOOD RAW MATERIALS

Carbohydrates	*Polysaccharides:*	
	Starch:	Soluble starch
		Granular starch
		Resistant starch
	Non-starch polysaccharides:	
	Cellulose	
	Pectins	
	Hemicelluloses:	Arabinoxylans
		$(1 \rightarrow 3, 1 \rightarrow 4)$-$\beta$-Glucans
		(Arabino)Galactans
	Oligosaccharides	
	Monosaccharides	
Proteinaceous material	*Proteins*	Globular proteins
		Fibrillar proteins
		Others
	Peptides	
	Amino acids	
Lipids	*Saponifiable lipids*	Triglycerides
		Phospholipids
		Glycolipids
	Non-saponifiable lipids	Free fatty acids[1]
		Sterols
		Waxes
Minor components	*Miscellaneous*	Anti-nutritional
		components,
		Minerals, etc.

[1] Free fatty acids are classified chemically as non-saponifiable lipids. In the practice of lipid analysis in feedstuffs, however, free fatty acids appear in the (alkaline) aqueous phase, and thus are analytically classified as part of the 'saponifiable' fraction.

weight polysaccharides according to the number of individual mono-saccharides from which the carbohydrate is built.

Monosaccharides

Monosaccharides can be classified on the basis of the number of carbon atoms. In food and feed raw materials, pentoses (five carbon atoms) and hexoses (six carbon atoms) are commonly present. Monosaccharides differ in the chiral position of the different hydroxyl groups and the position of the carbonyl group (aldehyde or keto). All monosaccharides are readily soluble in water and are present in a cyclic form via the formation of a semi-acetal. Monosaccharides can be oxidized in reactions with mild oxidants.

Oligosaccharides

Oligosaccharides are made up of a limited number (up to 10) of mono-saccharides coupled by glycosidic linkages. From a chemical point of view they can be divided into reducing (e.g. maltose, lactose) and non-reducing oligosaccharides (e.g. saccharose, raffinose) depending on the presence of a reducing end group. Most of oligosaccharides present in feed and food raw materials are readily water-soluble.

Polysaccharides

Whereas from a scientific point of view polysaccharides should be classified on the basis of the constituent monosaccharide composition, in practice various classification criteria are used like monosaccharide composition, glycosidic linkage type, morphological appearance or solubility of the polysaccharide. A first classification of polysaccharides is the division into starch and non-starch polysaccharides.

Starch is made up of two glucose polymers: amylose, a linear chain of glucose units interlinked via α-1,4 glycosidic linkages and amylopectin, a branched glucose polymer in which glucose monomers are interlinked via α-1,4 (in the chains) and α-1,6 (at branching points) glycosidic linkages. In plant material starch is mostly present in granules in which the two types of polymers are associated by intramolecular forces. Starch granules are insoluble in water below 50°C but upon mechanical treatments part of the granules can be partly disintegrated leading to solubilized molecules. In addition, upon moist heat treatment the starch granules can swell to many times their original size and upon continuous heating the granules start to disintegrate leading to solubilization of the individual starch molecules.

In both native and processed material resistant starch can be present. Resistant starch can be defined as that part of the total starch present which can not be digested in the upper part of the digestive tract and is fermented in the colon as dietary fibre. Resistant starch can be formed by recrystallization of solubilized amylose. In addition, native starch with a so-called B-type crystallization pattern and starch granules embedded in a cell wall matrix can also not be digested.

Non-starch polysaccharides (NSP) can be identified as those polymeric carbohydrates which are different in composition and structure (glycosidic linkage type) from amylose and amylopectin. These polysaccharides can be located intracellularly (e.g. fructans, mannans) or extracellularly (e.g. exudates) but the majority of the non-starch polysaccharides originate from the plant cell wall (e.g. arabinoxylans, cellulose, mannans). The latter are integral parts of cell wall structures together with proteins and often with lignin. These complexes are not readily accessible for enzymes in general and not degradable by the digestive enzymes of monogastric animals in particular. In the colon of these animals they are fermented to free fatty acids which are subsequently absorbed.

Based on their solubility in different solvents they can be classified into pectins (soluble in hot aqueous solutions of chelating agents or weak acids), hemicellulose (soluble in alkali) and cellulose (only soluble in concentrated acids). Whereas the cellulose fraction comprises only one polymer, a β-1,4 linked glucose polymer, both the pectin and the hemicellulose fraction comprise various different polysaccharides either chemically interlinked or not. Depending on the source of feed raw material the monosaccharide composition of the non-starch polysaccharides can differ greatly. In by-products of the fruit industry pectin is the most abundant group of molecules whereas in cereal (by-)products arabinoxylans (hemicelluloses) and cellulose are predominant.

The properties (e.g. solubility, viscosity, water-binding capacity) of the non-starch polysaccharides strongly depend on their structure and large differences exist between the pectin and hemicellulose fractions from different origins. Therefore, characterization and quantification of non-starch polysaccharides in terms of hemicellulose, pectin and cellulose, or even into the commonly used classes neutral-detergent and acid-detergent fibre (Van Soest, 1976), is not sufficient for understanding the behaviour of these components in feed raw material processing.

Proteinaceous material

Amino acids

Amino acids are characterized by the presence of at least one amino group and one carboxylic group within their molecular structure. Whereas in nature over 200 different amino acids have been identified only a limited number is widely present in food and feed raw materials.

From a nutritional point of view amino acids are divided into essential and non-essential amino acids. With respect to their chemical properties and reactivity the most common amino acids can be classified into 1) amino acids with nonpolar, uncharged (hydrophobic) side chains, 2) amino acids with uncharged polar (hydrophilic) side chains and 3) amino acids with charged (at a pH close to 7) side chains. A further subdivision is based on the presence of aromatic groups and the type of polar and ionic group present.

Peptides

Peptides are made up of a limited number of amino acids interlinked via peptide bonds. In the peptide bond the carboxyl and the amino group of two different amino acids participate. Although there are no clear rules, chains of up to 20 amino acids long are often called peptides, whereas chains longer than 20 amino acids are usually called polypeptides or proteins.

Proteins

Proteins are mostly built up of a large number of 20 different amino acids, which are all in the so called L-configuration, interlinked by peptide bonds. In contrast to polysaccharides there is no classification based on the constituent monomers as the 20 amino acids are present in (nearly) all different proteins. They can be present as individual polypeptide chains or as two or more single polypeptide chains associated by non-covalent forces or by disulphide bridges. Single polypeptide chains can vary in size from 20 to 500,000 amino acid residues.

Based on their properties and structure proteins can be divided into fibrillar (fibre-like form) and globular proteins, with some exceptions like caseins (more random coil like) and glutenins. Globular proteins have a fairly uniform amino acid profile with a relatively high proportion of acidic/basic amino acids. Fibrous proteins have high proportions of relatively few amino acids. e.g. glycine, proline (Franks, 1988)

Proteins can also be divided on the basis of their solubility in water, salt solutions and aqueous ethanol solutions and acid/alkaline solutions. Based on a sequential extraction procedure with the above mentioned solvents proteins can be divided into albumins, globulins, prolamins and glutelins, respectively.

The amino acid sequence of proteins, together with the physical structure (e.g. random coil, globular proteins, fibrillar) in which the proteins occur, determines the enzymic degradability and/or digestibility of proteins. Upon exposure to moist heat the physical structure usually changes causing substantial differences in reactivity, functional and nutritional properties.

Lipids

Lipids are a group of biomolecules with a pronounced hydrophobicity. By definition, lipids are water insoluble but very soluble in apolar organic solvents. In practice most of the lipids are derivatives of fatty acids. Lipids can be classified according to this presence of a fatty acid residue esterified to another group: acyl or saponifiable lipids and simple or non-saponifiable lipids, which are either free fatty acids or do not have an acyl group (Belitz and Grosch, 1987). As is stated in Table 5.1, it should be kept in mind that in the practice of lipid analysis there are discrepancies with the above classification.

In Table 5.1 the major examples of the two classes of lipids can be seen. Lipids can also be classified according to other characteristics. Some of the lipids (e.g. triglycerides, most free fatty acids, sterols) are neutral while others contain both an hydrophobic and hydrophilic group in their structure and are called polar or amphiphilic lipids (e.g. phospholipids, glycolipids). In the acyl lipids, which form the majority of all lipids found in food and feed products, the acyl or fatty acid group can differ in length and in the presence of one or more unsaturated bonds in the chain. This presence of an unsaturated bond or in particular the presence of two or more conjugated unsaturated bands makes the fatty acid prone to oxidation.

Minor components

Compound feed also contains many 'minor components' (or minor constituents), present in only grams/ton quantities. These components are usually denoted by their functional and physiological properties, but partly also by chemical structure.

One may distinguish 'true' minor components, belonging to classes of chemical structures that each are only present in feed in minor quantities, e.g.: vitamins, pigments, flavours, minerals (in particular trace elements), phytate, glycosinolates, others (incl. reaction products formed in processing) and 'pseudo' minor components, compounds with unique functional or physiological properties, but chemically belonging to the above mentioned classes of major constituents present in animal feed. Some minor constituents, e.g. protease inhibitors, lectins, antigenic proteins, enzymes are proteins.

Minor components may originate from the raw materials used, but may also have been added to meet nutritional or functional requirements. The ones with essential nutritional value are usually also added to the compound feed, before or after the thermo-mechanical processing, in order to compensate for losses caused by the processing. On the other hand, the aim of many (hydro-)thermo-mechanical processes is to inactivate minor components with undesired (anti-nutritional) properties.

REACTIVITY OF FOOD COMPONENTS

In Table 5.2 an overview of some of the most likely reactions of the three major feed components (carbohydrates, proteinaceous material and lipids) is given. The interest of the nutritionist is largely focused on those which are important for the nutritional quality of the feed, while the technologist is also interested in minor component that determine 'in process' behaviour functional properties of compound feed (e.g. binders).

Table 5.2 REACTIVITY OF MAJOR FOOD/FEED COMPONENTS

Carbohydrates	Proteinaceous material	Lipids
Maillard reaction	Maillard reaction	(Auto)oxidation
Caramelization	LAL/LAN formation	Thermal degradation
Solubilization	Deamidation	Cis/trans isomerization
Resistant starch formation	D-amino acid formation	Polymerization
	Iso-peptide formation	Formation of Maillard
	Denaturation	reactants

Carbohydrates

During (moist) heat treatments the reducing sugars can take part in cara-melization and Maillard reactions resulting in flavour changes and loss of nutritional value. As a result of thermo-mechanical treatments part of the NSP can be solubilised from the cell wall matrix through degradation or loosening of physical binding forces. This can contribute to the viscosity of the chymous (e.g. arabinoxylans) or open up the cell wall structures for release of nutrients. As mentioned before, upon moist heat treatment starch granules can (swell) gelatinize and become more digestible. However, depending on the conditions, resistant starch can be formed which is not digestible by monogastric animals.

Proteinaceous material

Amino acids, peptides and proteins, the latter by the presence of the essential amino acid lysine, can, due to the presence of the amino group, take part in the Maillard reaction. By heat-treatment at alkaline conditions essential amino acids present in proteins can be cross-linked and upon hydrolysis of the proteins unusual compounds like lysinoalanine (LAL) and lanthionine (LAN) are formed which can not be digested. Also, under these conditions formation of non-nutritive D-amino acids from the naturally occurring L-configuration can take place. Heating proteins at neutral pH in (semi-)dry conditions can result in the formation of iso-peptide bonds (cross-linking). It may be obvious that in addition to those reactions which cause a direct lower nutritional quality the cross-linking reactions may significantly change the protein structure as well as the total feed matrix structure by which the nutritional quality can also change.

The above mentioned reactions cause a change in the chemical structure of the amino acids, peptides and/or proteins. For the proteins this can lead to denaturation. However, denaturation of proteins can also take place at elevated temperatures without these reactions occurring. The proteins will denature and can aggregate, by which their solubility generally decreases and their digestibility increases. Bio-active proteins like enzymes, lectins and trypsin inhibitors will be inactivated. Due to denaturation hydrophobic interactions can take place and thiol groups may be exposed for interaction.

Lipids

Lipids are liable to oxidation, particularly lipids which contain unsaturated fatty acids. Heat treatments on the one hand prevent enzymic oxidation by inactivation of lipoxygenase but, on the other hand, enhance chemical oxidation which will proceed even during storage at low temperatures, particularly when pro-oxidants are present in the feed. The products of the oxidation process can take part in the earlier mentioned Maillard reaction. Lipid oxidation products can give strong off flavours to the feed; they can also react with proteins and reduce their digestibility. Upon heat treatment isomerization of double bonds from the cis configuration to the trans configuration can take place. Also, due to cross-linking of acyl chains dimerization/polymerization can occur leading to increased viscosity.

Minor components

Some of the minor components are heat labile whereas for others heat treatments do not seem to have an important effect on their stability.

THERMO-MECHANICAL TREATMENTS (PROCESSES) IN THE ANIMAL FEED INDUSTRY

The major thermo-mechanical processes used in the production chain of animal feed are listed in Table 5.3. In the framework of this paper, thermo-mechanical treatments are looked at from a 'reactor approach'. The conditions during the process determine the reactivity of the various

Table 5.3   THERMO-MECHANICAL TREATMENTS IN THE ANIMAL FEED INDUSTRY

Process	Temp. [°C]	Moisture content[3]	Resistance time	Shear
Toasting[1]	100–140	low	minutes	-
Drying[1]	>100	medium	minutes	-
Steam flaking	≈100	medium	minutes	+
Steam explosion[1]	140–210	medium	seconds	++
Preconditioning[1]	<100	medium	minutes	-
Grinding[1,2]	≈20	low	seconds	++
Granulation[2]	50–95	medium	minutes	-
Pelleting[2]	60–100	low	seconds	+
Expansion[2]	80–140	medium	seconds	+
Extrusion[2]	90–140	medium	seconds	++

[1] used as pretreatment; [2] used as main process; [3] low ≤ 18%; medium: 18–30%

chemical entities contained in the materials, the type of reactions and the extent to which they will occur.

The major parameters characterising the process are temperature, moisture content, or more correctly water-activity (A_w), residence time and shear. The ranges given in Table 5.3 for each parameter should be seen as an indication only. For example, the type of shear created in a hammermill, an expander or during steam explosion, are each of a very different nature. The true conditions vary widely in practice. In addition, the effect of the process is also influenced by parameters determined by the feedstock used, and process conditions like pH and pO_2.

Some of the processes to which feed raw materials are exposed are carried out by the industries involved in the primary processing of agricultural products. Products for the feed industry are in many cases byproducts, emerging from processes with different objectives. The thermomechanical processes are often applied in order to achieve certain specific goals such as gelatination of starch, inactivation of endogenous enzymes and trypsin inhibitors, drying or product shaping. Other effects of the process such as the formation of new substances or the loss of essential elements have to be accepted or are neglected as long as their impact on product properties is acceptable. The consequences for shelf-life, further processing and final functional and nutritional properties of animal feed products may be substantial however.

From Table 5.3 it can be concluded that in most cases the temperature/time combinations are sufficient for various of the reactions mentioned in Table 5.2 to occur. The moisture content during the reactions is often low or intermediate, being not favourable for most of the reactions to occur. However, as the water-activity and not the moisture content is important and depending on the product the water-activity can be sufficient high (or low) enough to cause the above mentioned reactions.

Maillard reaction

OUTLINE OF THE MAILLARD REACTION

For the carbohydrate components one of the most important reactions is the Maillard reaction in which many constituents of raw materials can participate and which affects many quality attributes (Figure 5.1). Reactions between reducing sugars and free amino groups from amino acids, peptides and proteins prevail, but oxidised vitamin C also is very reactive with reducing compounds formed by oxidation of lipids and polyphenols.

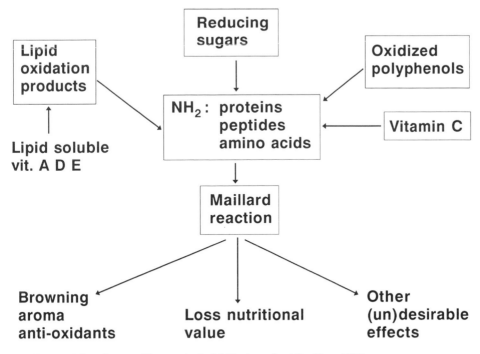

Figure 5.1 Constituents taking part in the Maillard reaction (after Kato, 1987).

The involvement of vitamin B_1 (thiamine), through its amino group, and vitamin B_6 (pyridoxine), through its aldehyde group, in Maillard reactions has also been shown, including loss of pantothenic acid and vitamin B_{12} (cobalamin) on prolonged storage of milkpowder (Hurrell, 1990). In food processing Maillard reactions are often essential for the formation of characteristic colour and flavour, e.g. in baking, cooking, roasting. However, there are also important negative aspects to Maillard reactions such as loss of nutritional value by reduction in essential amino acids, and in reduction of digestibility of proteins through cross-linking reactions, loss of nutrients, and reduction in bio-availability of trace elements like iron and zinc and in extreme conditions the formation of toxic compounds. Processes in food manufacture are therefore optimized with respect to Maillard reactions by the choice of raw materials, ingredients, moisture content and of the type of process and process conditions including packaging and storage conditions. The situation in feed manufacture is more complicated: some raw materials are often already heavy 'maillardized' or rich in compounds susceptible to Maillard reactions in subsequent thermal processing under conditions favourable for Maillard reactions.

The scheme proposed by Hodge in 1953 (Hodge, 1953) to describe the many reaction pathways of the Maillard reactions is still considered the most satisfactory representation (O'Brien and Labuza, 1994). Figure 5.2 gives a simplified overview of Maillard reaction pathways. Three stages are distinguished: an initial stage, an intermediate stage and an end stage. In the initial stage a nucleophilic addition of the amino group of an amino compound to the electrophilic carbonyl groups of a reducing sugar or another carbonyl compound takes place and a glycosylamine is formed. This is still a reversible reaction which is followed by a re-arrangement of the glycosylamine to so called Amadori (aminoketose) or Heyns (aminose) compounds. These latter compounds are rather stable and once formed the reaction is irreversible. Amadori and Heyns compounds can be analysed easily and they are found to be present in heat treated, dried and stored foodstuffs.

In the intermediate stage Amadori and Heyns compounds are further degraded by a myriad of very complex reactions. Depending on the conditions three main routes may be followed; one route starts with 1-2 enolisation followed by elimination of water and the amino group giving, among others, 3-deoxyosons (reactive dicarbonyls), furfural or hydroxy-methylfurfural (HMF). The presence of these latter compounds in a food is often used as an indicator for heat damage. These reactions take place particularly at pH less than 3. Another route, favoured by a pH greater than 7, begins with 2-3 enolization, followed by deamination and the formation

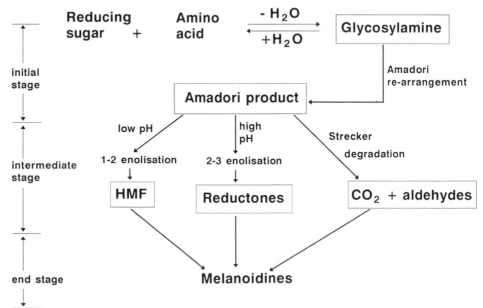

Figure 5.2 Simplified scheme of Maillard reaction pathways.

of 1-deoxyosons which subsequently fragment into many small, volatile flavour components. Many of these fragmentation products have strong reducing carbonyl groups which engage in new reactions with amino compounds. In products with a high content of amino acids and treated at high temperatures ($> 120°C$) amino acids react with dicarbonyl compounds by the Strecker degradation mechanism. After transamination and decarboxylation CO_2 and aldehydes are formed next to pyrazines, oxazoles, and, from S-containing amino acids thiazoles. The changes taking place in this stage are not yet visible; in model systems they are followed by measuring the increase in absorbance at 420 nm. By sniffing, formation of aromatic compounds can be detected.

In the end stage many of the reaction products formed in the preceding reactions condense and polymerise to high molecular, brown complexes called melanoidins. Amino groups are strongly involved in these reactions noticeable from the fact that in this stage a strong reduction in available (essential) amino acids is observed.

In humans Amadori compounds have an absorption of about 60% in the intestines and are excreted unchanged in the urine, 40% is fermented in the colon by which the amino acids are set free and partly absorbed. HMF is excreted as furfurylic acid in the urine. The melanoidins are not absorbed. They are active in the digestive tract as inhibitors of digestive enzymes and decrease the bio-availability of minerals like iron and zinc.

FACTORS AND CONDITIONS DETERMINING MAILLARD REACTIONS

To be reactive in Maillard reactions, sugars need an aldehyde or keto group. Sugar alcohols are not reactive and sucrose is reactive only after hydrolysis to glucose and fructose. In general pentoses are more reactive then hexoses, aldoses more than ketoses and monosaccharides more than di- and oligosaccharides. Dehydro-ascorbic acid, which can be considered as a derivative of L-gulonic acid is very reactive and also uronic acids like galacturonic acids are important. The reactivity of amino compounds depends on the concentration of free amino groups. Amides are therefore more reactive than amino acids and basic amino acids more than neutral amino acids. Neutral amino acids are in general more reactive than acidic and polar amino acids. In proteins only basic proteins have a free amino group. This is the reason that particularly the ε-amino group of lysine is very reactive.

The influence of temperature on Maillard reactions is very complex.

With increasing temperature the reaction velocities increase strongly (Q_{10} ca 2.5) and the reaction routes also change. At low temperature (< 60°C) Amadori compounds are formed and there is hardly browning. Hurrell, Finot and Ford, (1983) reported a decrease in available lysine of 30% and a parallel formation of lactusyllysine in milkpowder (relative humidity 2.5%) stored at 60°C. At temperatures over 100°C Strecker degradation and melanoidin formation prevail. The types of reactions taking place are also influenced by pH and moisture content of the product. The effect of temperature on formation of HMF in heated skim milk is shown in Figure 5.3 (Berg, 1993).

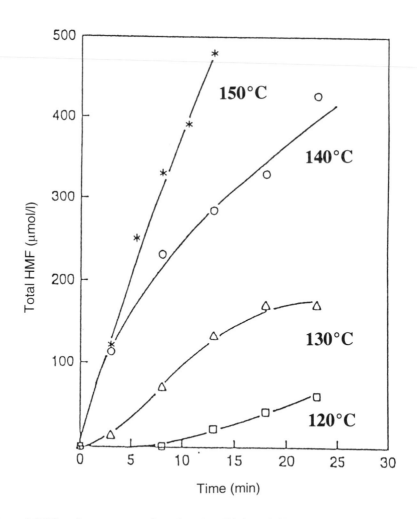

Figure 5.3 Effect of temperature on formation of HMF in heated skim milk (after Berg, 1993).

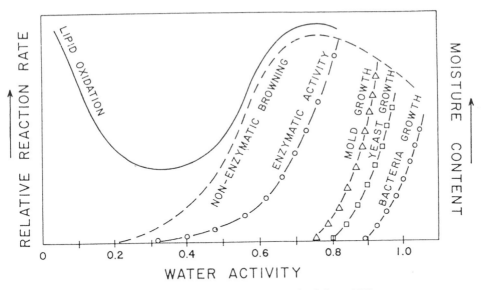

Figure 5.4 The effect of water-activity on different reactions (after Labuza, 1971).

Maillard reactions are minimal in the pH range 3-4. Lower pHs can cause formation of reducing sugars and amino acids by hydrolysis (sucrose) and thus promote Maillard reactions. From pH 4 Maillard reactions increase to an optimum at pH 10, at higher pH values Maillard reactions decrease. The effect of water-activity on Maillard reactions (non-enzymic browning) is shown in Figure 5.4, they start off at an A_w of 0.25 (ca. 18% moisture content) and increase to an optimum at an A_w of ca. 0.7 (moisture content 30-35%). At higher A_w a small decrease can be observed.

Maillard reactions can be limited by operating in a pH range of 3-4, at a low moisture content, by keeping the process temperatures as low as possible or by carrying out high temperature treatments in a very short time (HTST process). Also combinations of these factors can be effective. Addition of sulphite (SO_2) strongly reduces reactions in the end stage of the Maillard reaction. It reacts with carbonyl-intermediates and prevents condensation and polymerization reactions so that no brown melanoidins are formed. Early Maillard reactions and Strecker degradation are, however, not inhibited by sulphite. Another way to minimize Maillard reactions is to remove or lower the concentration of reactants, for instance by fermentation of reducing sugars.

Protein denaturation

DENATURATION PROCESS

Thermo-mechanical treatments are often used to increase the nutritional value of proteins by inactivation of antinutritional factors (ANFs) and protein denaturation. Inactivation of ANFs only is not sufficient to explain the increase in protein digestibility which was obtained after thermo-mechanical treatments (Marsman, Gruppen and van der Poel, 1993; Liener, 1994). This suggests that denaturation of storage proteins is also an important process in thermo-mechanical treatments.

Protein denaturation can be defined as any change in the conformation of a protein that does not involve the breaking of peptide bonds (Camire, Camire and Krumhar, 1990). Most proteins undergo structural unfolding followed by aggregation when subjected to moist heat or shear. Unfolding is usually a reversible process and if the thermo-mechanical treatment is stopped before aggregation begins, the protein can return to its native conformation (Figure 5.5). If more heat or shear is added non-covalent interactions like hydrophobic and electrostatic interactions, which contribute to the stabilization of the three-dimensional structure, will be broken, resulting in irreversible protein denaturation (Figure 5.5). Thermo-mechanical treatment may also result in the breaking of covalent bonds such as disulphide bonds. During denaturation hydrophobic groups are uncovered resulting in a decreased solubility of the proteins in aqueous

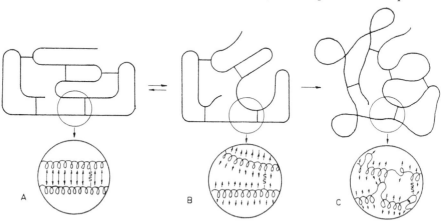

A: native state
B: reversible denatured
C: irreversible denatured

Figure 5.5 Changes in protein structure due to denaturation (from Jaenicke, 1965).

solutions. In the literature the process of unfolding and aggregation is often combined in to one parameter called protein denaturation.

If during processing the specific mechanical energy (SME) is high enough, protein denaturation can be followed by association and dissociation reactions and also by breaking or formation of some covalent bonds e.g. hydrolysis of peptide bonds, modification of amino side chains (lysinoalanine or Maillard products) and the formation of new covalent isopeptide cross-links (Stanley, 1989). Upon cooling large protein complexes can be formed by inter- and intra-molecular interactions.

FACTORS DETERMINING PROTEIN DENATURATION

From Table 5.3 it can be concluded that temperature, moisture content and shear forces are the main factors influencing the denaturation process. To a lesser extent the residence time, pH and the presence of other components are of importance. Parameters like moisture content, temperature and shear forces are highly interrelated (Van Zuilichem, Jager and Stolp, 1988).

Moisture content, temperature and shear forces

Proteinaceous rich materials behave like non-Newtonian fluids under process conditions. Therefore, viscosity is observed to be highly dependent on moisture content, temperature and shear rate (especially during extrusion, expansion and pelleting). During extrusion or expansion, mechanical dissipation of the energy input as friction of the particles in the extruder/expander channel will result in an increase in the product temperature. If the initial moisture content of the raw material is increased, the viscosity and therefore the viscous dissipation will decrease resulting in lower product temperatures (Marsman *et al.*, 1995). In those processes it is almost impossible to change one of the parameters while keeping the other process conditions constant.

The denaturation temperatures (T_d) of most proteins in solutions are usually below 100°C. However, the moisture content during processing, which is generally lower than 30%, affects the T_d. Since differential scanning calorimetry (DSC) has proved to be a valuable tool in the analysis and finger-printing of a wide array of proteins and proteinaceous systems, it was applied to the study of protein denaturation. With this method the heat needed for denaturation of proteins, which is an endothermic process, can be measured. From a study of Kitabatake and Doi (1992) the T_d of the main storage proteins in soybean meal, conglycinin (7S) and glycinin (11S), appeared to be 76.5 and 93.3°C, respectively at a moisture content of 94% (Figure 5.6). If the moisture content decreased to 29% or less no T_d of glycinin could be found and the T_d of the conglycinin shifted to a temperature exceeding 180°C. In another study toasting of soya flour at 137°C only caused a partial (about 40%) denaturation (Jurgens, Van Maanen, Vooijs and Beumer, 1993). Proteins become heat stable when the moisture content is lower than 30%. The lower the moisture content the higher the T_d (Biliaderis, 1983). It is expected that additional shear forces during pro-

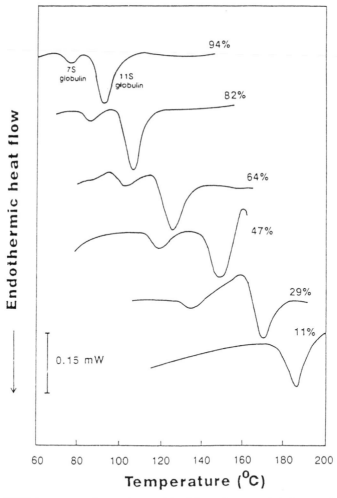

Figure 5.6 DSC thermograms for samples containing 3.5 mg of soya protein isolate at various moisture contents (from Kitabatake and Doi, 1992).

cessing can lower the T_d compared with processes in which no shear forces are involved.

Residence time distribution

Residence time is a process parameter which can be relatively easily adjusted during most thermo-mechanical treatments. For extrusion, expansion, steam explosion, pelleting and grinding the residence time is in the range of seconds, while treatments like toasting, drying, granulation and preconditioning are processes in which the residence is mostly expressed in minutes. More important than the average residence time, in some processes, is the residence time distribution (RTD). A high RTD means that parts of the proteins are exposed for a longer time to heat and shear resulting in deviated chemical, physical and physiological behaviour. The RTD can be measured, but it is a result of all mass flows present during processing (Jager, Zuilichem, Stolp and van't Riet, 1989).

pH

The pH also has an influence on protein denaturation. Greater thermo-stability is seen in the isoelectric region (pH 4-5) where the net charge of the proteins is low. As one moves away from the pH region (pH $<$ 4 or pH $>$ 9) the T_d decreases considerably. These extreme pH values are not favourable in the feed industry, because of unwanted reactions like deamination of proteins and the Maillard reaction (Stanley, 1989).

The influence of other components on protein denaturation

Protein denaturation may also be influenced by the presence of other components. Besides the negative effects of auto-oxidation, thermal degradation and polymerisation that lipids can exhibit during processing, they also prevent the expansion of proteins, which is necessary for a good texturized product (Cheftel, 1986). The denaturation of proteins is also inhibited by the presence of carbohydrates. It may be that, in some way, the embedded carbohydrates effectively stabilize the more labile hydrophobic interactions between proteins or gives rise to additional stabilizing forces (Sheard et al., 1984). In the presence of carbohydrates extrudates are more resistant to stress and the protein matrix is less disrupted (Camire et al.,

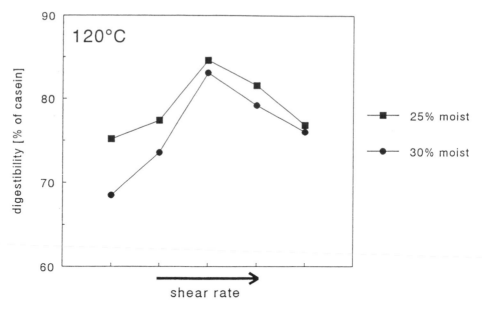

Figure 5.7 *In vitro* protein digestibility (% of casein standard) of untoasted soybean meal extruded at 120°C and at different moisture contents and shear levels (from Marsman *et al.*, 1995).

1990). However, as mentioned above, reducing sugars can cause Maillard reactions. The process conditions used in thermo-mechanical treatments, such as a high temperature and a low moisture content, are known to favour the Maillard reaction. Clearly, the interaction with other food components during processing on the protein denaturation merits further research.

MONITORING EFFECT OF THERMO-MECHANICAL TREATMENT

Proper processing of raw materials includes a precise control of temperature, residence time, moisture content and shear forces. Adequate analytical methods are necessary to monitor the chemical, physical and physiological changes resulting from processing. Optimal processing means a destruction of ANFs and, if possible, optimal denaturation of proteins. Under- and over-processing can result in a product of lower nutritional quality. Under-processing results in incomplete inactivation of heat labile ANFs, while over-processing may result in a reduced availability of lysine.

In vitro protein digestibility

The most important parameter in evaluating protein quality as a result of processing is the *in vivo* nutritional value. Different kinds of *in vitro* methods have been developed to predict this. A rapid multi-enzyme technique for estimating the *in vitro* protein digestibility was suggested by Hsu, Vavak, Satterlee and Miller, (1977). They measured the pH drop after adding proteolytic enzymes to samples containing the processed proteins. The results they obtained were closely related to the *in vivo* digestibility. Later, this method was modified by keeping the pH constant during enzyme incubation. The amount of sodium hydroxide needed to neutralize the production of hydrogen ions, was an indication of the *in vitro* protein digestibility (Pederson and Eggum, 1983). This pH-STAT method has also been used in experiments in our department. Untoasted soyabean meal, extruded at 120°C at different shear rates and initial moisture contents, showed an increase in the pH-STAT digestibility if the shear rate was increased (Figure 5.7). After a maximum, higher shear rates resulted in a decline of the pH-STAT digestibility. Most likely, the increase in the *in vitro* protein digestibility could be explained by inactivation of trypsin inhibitors. However, in a previous study it was concluded that extrusion of untoasted soyabean meal at 120° and at a low shear input already decreased the trypsin inhibitor by more than 90% to values of about 2 mg/g sample, which is considered to be low enough in the feed industry (Marsman *et al.*, 1993). This would mean that the increase in pH-STAT digestibility with increasing shear levels cannot be ascribed to trypsin inhibitor inactivation. Shear forces are thought to play an important role in the formation of the final protein structure during extrusion. Increasing shear forces denatures proteins more easily and proteins are, therefore, more accessible for enzyme attack (Bhattacharya and Hanna, 1988). If energy is in excess, all kinds of cross-linking reactions of protein molecules can occur e.g. Maillard reaction, lysinoalanine formation or deamination, resulting in a decreased nutritional value. This is, probably, the reason why the pH-STAT digestibility decreased if the shear input was too high (Figure 5.7).

The relation between *in vitro* and *in vivo* protein digestibility has always been questioned. In the past some rapid screening methods were developed to monitor protein denaturation and thereby the nutritional value.

Urease activity

In the feed industry the urease activity, measured as a pH rise in an ammonia solution, has long been used to monitor the heat treatment of soya products. It is said that destruction of urease parallels that of trypsin inhibitor inactivation (Waldroup and Smith, 1989). While the urease activity is a satisfactory measure of heat treatment of soya products up to the

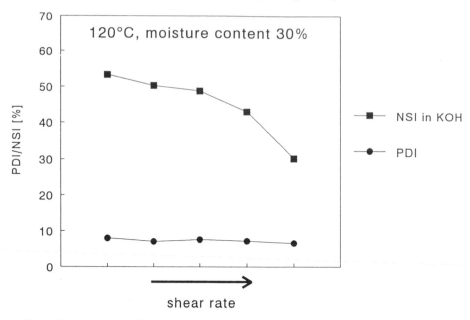

Figure 5.8 Protein dispersibility index (PDI) and nitrogen solubility index (NSI) in potassium hydroxide (%) as a result of extrusion of untoasted soyabean meal at a temperature of 120°C, initial moisture content of 30% and at different shear levels (from Marsman *et al.*, 1995).

point of optimum heating, it is of no value in detecting possible over-processing.

Protein dispersibility index

As stated before, protein denaturation is accompanied by a decrease in protein solubility. By measuring the protein solubility one can get a good impression as to what extent proteins are denatured. The protein dispersibility index (PDI) is often used to characterise raw and processed materials. This method, also called the 'fast stir' method measures the proportion of protein which is dispersible in water. Untreated raw materials have PDIs of about 80%, while thermo-mechanically treated materials have PDIs of 20–40%. Untoasted soyabean meal with a PDI of 80% and an initial moisture content of 30% was extruded at 120°C on a single-screw extruder. The influence of shear forces on protein solubility was studied by using torpedo elements of different lengths (Marsman, Gruppen, van Zuilichem, Resink and Voragen, 1995). It appeared that in this study the PDI dropped to levels below 10% after extrusion at a low shear rate and did not change if the shear rate was increased (Figure 5.8). The same results were found for extrusion of soyabean meal (Hendriks, Moughan, Boer and van der Poel, 1994) and spray drying, toasting and extrusion of defatted soya products (Visser and Tolman, 1993). However, the value of the PDI seems to depend on the type of material. In peas, the PDI was a much better indicator in evaluating protein quality after processing (van der Poel, Stolp and Zuilichem, 1992).

Nitrogen solubility index in potassium hydroxide

In quality programmes in the American poultry industry, an alternative method of evaluating the protein quality as a result of processing has been used frequently (Dale *et al.*, 1987). In this method the proteins are extracted in alkaline solutions using 0.042M potassium hydroxide. Several studies have been performed to evaluate the usefulness of nitrogen solubility index (NSI) in potassium hydroxide. After autoclaving soyabean meal at 121°C it was concluded that NSI values below 70% depressed the growth of broiler chickens, but NSI values in excess of 85% indicated that soyabean meal was under-processed (Araba and Dale, 1990). In a study on the effect of shear forces during extrusion of untoasted soyabean meal (NSI 90) at 120°C, it appeared that, if the shear rate was increased, the NSI declined, especially at high shear rates (Figure 5.8). Comparing these results with other studies, it was concluded that determination of the NSI in soyabean meal processing could be a better indicator than the PDI, but that every process has its own NSI range (Marsman *et al.*, 1995).

NEW TRENDS IN MONITORING THERMO-CHEMICAL TREATMENTS IN MORE DETAIL

Both NSI and *in vitro* protein digestibility are methods giving a different but overall view of the effects of processing on protein properties. However, these methods are inadequate to study the different types of interactions in proteins as a result of processing. In the food and feed industry there is a growing interest in studying the influence of thermo-mechanical treatments in more detail.

Solubilization measurements in monitoring protein-protein interactions

One topic is the study of the different types of interactions between proteins and how they are affected as a result of processing. The interactions involved in the process of folding and formation of proteins are covalent and non-covalent. Non-covalent forces, which stabilize the native conformation of proteins and therefore influence their functional behaviour, are:

- hydrophobic interactions
- electrostatic interactions
     hydrogen bonds
     dipole-dipole interactions
- van der Waals' interactions

The strengths of these interactions are relatively small, especially the dipole-dipole and the van der Waals' interactions, but the non-covalent interactions are numerous and therefore these interactions can play an important role in the determination of protein structure. The most important covalent interaction is the disulphide bridge between two cysteine amino acids. The number of these bridges is relatively small, but more energy is needed to cleave this type of linkage than other types of interactions.

During thermo-mechanical treatments, proteins are unfolded and thereby covalent and non-covalent interactions are broken. The process of denaturation can be followed by aggregation, association and breaking and formation of new peptide bonds. One approach for studying protein-protein interactions is based on protein solubilization. Reagents with a well known mode of action are used. Addition of urea will cause gross solubilization of denatured molecules and small aggregates held together by hydrophobic and hydrogen bonds. Sodium sulphite or dithiothreitol (DTT) can be added to cleave disulphide bonds that produce larger aggregates insoluble even in urea (Areâs, 1992).

Hager (1984) applied this method to characterise extruded soyabean concentrate. This study provided evidence that intermolecular disulphide bonding is an important factor contributing to extrudate structure, at least for extrusion temperatures below 150°C. Around temperatures of 140°C, the activation energy for peptide bonds was not attained (Hager, 1984). In another study on the protein interactions in soya processing it is claimed that disulphide bonds were of negligible importance, suggesting that new peptide bonds were formed under severe process conditions (Burgess and Stanley, 1976). Studies on soya extrusion carried out at 140 to 180°C, however, showed that at all tested temperatures, disulphide bridges, followed by non-covalent interactions, were the prevalent type of protein-protein interactions on the extrudates (Areâs, 1992). Recently, the authors have studied how proteins were affected in soyabean meal as a result of toasting (20 minutes, 90°C) and extrusion (35 seconds, 120°C and initial

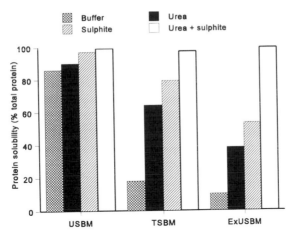

Figure 5.9 Protein solubility (%) after extraction of proteins from untoasted, toasted and extruded untoasted soyabean meal (USBM, TSBM and ExUSBM, respectively) in buffer, urea, sodium sulphite and urea and sodium sulphite.

moisture content of 30%) using the method described by Hager (1984). The results are given in Figure 5.9. It can be seen that in untoasted soyabean meal more than 80% of the proteins were already soluble in aqueous buffer (pH = 7.6). Addition of urea and/or sodium sulphite increased the solubility to 100%. In toasted soyabean meal protein solubility in the buffer was reduced compared to untoasted soyabean meal, suggesting protein denaturation and to a certain extent aggregation of protein molecules. Addition of urea increased the protein solubility to 60% and addition of sodium sulphite to 75%. Only a combination of the two reagents resulted in an almost complete solubilization of proteins. If untoasted soyabean meal was extruded, the protein solubility in urea was 35% and in sodium sulphite 50%. Most likely, proteins are more aggregated after extrusion compared with toasting. But also after extrusion a combination of sulphite and urea resulted in a complete solubilization of the proteins. In this experiment, extrusion had more impact on protein structure than toasting and it can be stated that non-covalent as well as covalent interactions play an important role during extrusion of soyabean meal.

Oxidation and thermal degradation of lipids

OUTLINE OF THE REACTIONS

Oxidation and thermal degradation of lipids lead to complex chemical changes and interactions of reaction products with other feed constituents

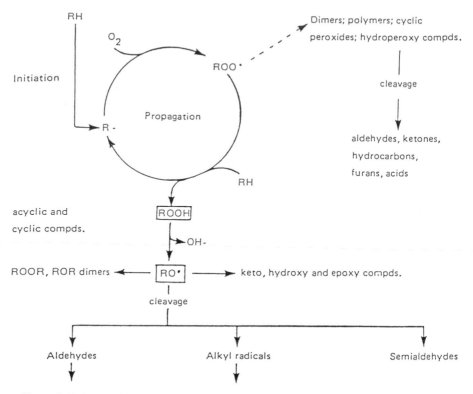

Figure 5.10 A generalized scheme for auto-oxidation of lipids (from Nawar, 1985).

producing numerous compounds deleterious to the quality and nutritional value of the feed. Typical for lipid oxidation is the formation of off-flavours generally called 'rancid', 'fishy' or 'cardboard-like'. A generalized scheme for (auto)oxidation of lipids is shown in Figure 5.10. The oxidation process is a radical-induced chain-reaction. The initiation step of auto-oxidation may take place by hydroperoxide decomposition, by metal ion catalysis or by exposure to light. Also, the involvement of singlet oxygen sensitized by plant or tissue pigments such as chlorophyll or myoglobin has been indicated. Upon formation of sufficient free radicals, the chain reaction is propagated by the abstraction of hydrogen atoms at positions α to the double bonds. Next, oxygen addition takes place at these positions and free peroxy radicals (ROO·) are formed. These radicals can in turn abstract hydrogen from α-methylenic groups RH of other molecules to yield hydroperoxides ROOH and R· groups. The new R· groups react with oxygen and the cycle can start again. Hydroperoxides start to decompose as soon as they are formed, first to an alkoxy and a hydroxy radical followed by carbon-carbon bond cleavage on either side of the alkoxy group. Cleavage on the carboxyl or ester side results in the formation of an aldehyde and an acid. Scission on the hydrocarbon side produces a hydrocarbon and an oxoacid. In addition to the breakdown, all kinds of interaction mechanisms take place leading to the production of many compounds which react with other constituents in the system (Belitz and Grosch, 1987).

Also during thermal treatments at high temperatures many chemical changes of lipids can take place. Non-oxidative degradation of saturated fatty acids only takes place at temperatures well over 200°C. Typical reaction products are hydrocarbons, ketones, oxopropylesters, propene- and propanediol diester, diacetylglycerols, acrolein, CO and CO_2. In the presence of oxygen saturated fatty acids also undergo oxidation giving rise to degradation. The reaction starts with the formation of monohydroperox- ides which preferentially takes place at the methylene group at the α, β, or γ position. Reaction products from this degradation are carboxylic acids, alkanons, alkanals, lactones and hydrocarbons. When unsaturated fatty acids are heated in the absence of oxygen dimeric compounds are pro- duced. Decomposition starts at temperatures over 220°C. Unsaturated fatty acids are very susceptible to oxidative decomposition at elevated temperatures. The principal reaction pathways at elevated temperatures are practically the same as under low temperature conditions. As well as decomposition, dimerization and polymerization reactions can also take place in lipids by thermal as well as by oxidative mechanisms, resulting in a decrease in the iodine value and an increase in molecular weight, viscosity and refractive index (Nawar, 1985).

FACTORS AFFECTING LIPID OXIDATION

Lipids with a high content of unsaturated fatty acids are more liable to oxidation than lipids high in saturated fatty acids. Free fatty acids are more easily oxidized than fatty acids esterified to glycerol. Of the unsaturated fatty acids *cis* fatty acids are more susceptible to oxidation than *trans* fatty acids and conjugated fatty acids more than non-conjugated. Lipids with a natural distribution of fatty acids are more stable than lipids randomized by transesterification. For oxidation of lipids oxygen is required. At low oxygen pressure the rate of oxidation is proportional to the oxygen pressure. If the supply of oxygen is unlimited the rate of oxidation is independent of oxygen pressure. Also, surface area of the product is important since this deter- mines the accessibility for oxygen. For powdered products this effect is of course quite different than for oil-in-water emulsions in which oxygen diffusion rates are also of importance. Increasing temperature generally increases the rate of oxidation. Transition metal ions like cobalt, copper, iron and nickel are potent pro-oxidants which decrease the length of the induction period and increase the rate of oxidation. Trace metal ions originate from soil, metallic equipment, as natural constituents of raw materials and biological fluids of plant and animal origin. Anti-oxidants suppress lipid oxidation and for that reason they are added to the fat-containing raw materials and to feed. The effect of moisture content can be observed in Figure 5.4. In dried products with low moisture contents oxidation proceeds very rapidly but with increasing moisture content oxidation decreases to a minimum at an A_w of about 0.3. The protective effect of water in this stage is due to reduction in the catalytic activity of metal ions, to quenching of free radicals and by decreasing the excess of oxygen to the lipid fraction in the food. The oxidation rate increases when the moisture content is further increased to an A_w of about 0.8.

Some important changes as a result of lipid oxidation reactions affecting quality and nutritional value are summarized in Figure 5.11. Peroxides and their derivatives affect the flavour and odour of the feed. When the peroxides interact with vitamins the nutritional value and colour (carotene) may change. When they interact with protein, cross-linking reactions take

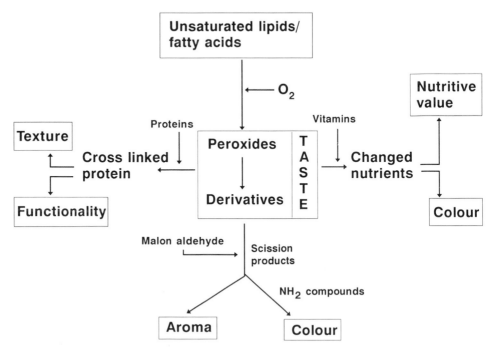

Figure 5.11 Effects of oxidation of unsaturated lipids on nutritional and quality aspects of foods (after Eriksson, 1982).

place which affect the texture, functional properties and, particularly, nutritional value. Scission products from peroxides contribute to aroma and after reaction with amino compounds also to colour formation. Bifunctional aldehydes can also cross-link proteins. In feeds containing heavily oxidized fats and oils, such as used frying oils, unfavourable effects originating from lipid degradation products should not be underestimated.

Reactivity of minor components

It is not the aim of this paper to discuss the impact of thermo-mechanical processes on minor constituents in detail. Their chemical structure and thus their reactivity under process conditions is very different.

Minerals are by nature quite stable, but they may act as catalysts and their bio-availability may change as a result of processing. The involvement of some minor components in the Maillard reactions and lipid oxidation has been mentioned before. In the inactivation of minor components with a protein structure, the mechanisms discussed for proteins apply. The extent to which inactivation takes place under the process conditions applied, will differ from compound to compound. The kinetics of inactivation of many 'Anti Nutritional Factors' has been reviewed in a recent symposium. (van der Poel, Huisman and Saini, 1993)

The stability of vitamins in hydro-thermal and thermo-mechanical processes depends on their individual chemical properties. Details can be obtained from the appropriate textbooks (e.g. Marks, 1975) and recent reviews (e.g. Flachowsky and Schubert, 1993). Recovery rates of the various vitamins may vary from almost 100% to less than 20%, depending on the process conditions and the formulation of the vitamin product used as feed supplement (Schai and Gadient, 1994).

Enzymes used in animal feed are denatured under moist heat and shear conditions, in particular when the temperature is above 75 °C. Special formulations may have a slightly improved resistance under not too extreme thermo-mechanical processing conditions. As a consequence of the rather high sensitivity of many essential minor components to rigorous thermo-mechanical processing, there is a tendency towards 'post processing' application of supplements containing these compounds, often in the liquid form.

Conclusions

In this paper, we dealt in particular with the effects of thermo-mechanical (and hydro-thermal) processing on the three major classes of chemical substances present in feedstuffs -carbohydrates, proteins and lipids. The intention was to advocate that food chemistry should be applied to inter-pret, predict and model processing technologies in the feed industry. In fact, the feed industry has to accept that a substantial part of its raw materials are provided by the agro-food industry, and may have been pre-processed in some way.

The properties of these materials are to some extent determined by this pre-processing, and properties may vary according to the processes applied. The feed industry should know however which properties can be expected as a result of certain pre-processing and might be able to calculate the properties of pre-processed products purchased from the agro-food indus-try. It should also have appropriate analytical methods available to charac-terize products and processes. In our opinion, the set of analytical methods commonly used in the feed industry is still insufficient to make these characterizations properly.

Within the scope of this paper, we could only deal in detail with the three most important types of reactions that occur in thermo-mechanical pro-cessing, the Maillard reaction, protein denaturation and lipid (auto) idation. In fact, many more biochemical, chemical and physical modifi-cations occur during processing and contribute to the changes of product properties caused by these processes. The impact of some highly complex cascades of reactions on the functional and nutritional properties of feed components was illustrated. The reactions and the effects are often not restricted to one class of compounds but may influence the nutritional availability of several major and minor compounds present in the complex feed mixture.

Reaction conditions in the pre-processing and final manufacturing of animal feedstuffs vary widely. Moisture content, an important determinant of reactivity, is sometimes high in pre-processing but often low to medium in final feed manufacturing. As was illustrated, water content has a different influence on various reactions. The effects of process conditions have to be estimated one by one for each process, using the general knowledge of the kinetics of the relevant reactions. Heterogeneity in raw materials, as well as in process conditions at 'micro' level, is a complicating factor.

One should be aware that some reactions (e.g. Maillard reactions, lipid oxidations), initiated by processing, continue during storage at ambient temperatures and may change properties of stored products substantially.

We have not dealt in detail with gelatinization of starches, inactivation of wanted or unwanted minor constituents and the elimination of micro-organisms, nor with the mechanisms of particle binding and shaping, often the major objectives for processing in the feed industry. However, much information on these subjects is already available in the feed industry. Losses of essential nutrients during processing are usually corrected for in the formulation by supplementation or increasingly by addition 'post processing'.

It is obvious that (present and future) economical constraints are limiting the possibilities to optimise thermo-mechanical processing in the feed industry. Many of the technologies and the appropriate analytical techniques for monitoring have been developed for the food industry. This knowledge can be made available to the feed industry to enable more optimal process designs and better cost/benefit optimisation in thermo-mechanical processing.

References

Araba, M. and Dale, N.M. (1990) Evaluation of protein solubility as an indicator of overprocessing soybean meal. *Poultry Science*, **69**, 76–83

Arêas, J.A.G. (1992) Extrusion of food proteins. *Critical Reviews in Food Science and Nutrition*, **32**, 365–392

Belitz, H.-D. and Grosch, W. (1987) *Food Chemistry*, Springer-Verlag, Berlin, Germany

Berg, H.E. (1993) *Reactions of lactose during heat treatment of milk: a quantitative study*. Ph.D. thesis. Wageningen Agricultural University, Wageningen, The Netherlands

Bhattacharya, M. and Hanna, M.A. (1988). Effects of lipids on the properties of extruded products. *Journal of Food Science*, **53**, 1230–1231

Biliaderis, C.G. (1983) Differential Scanning Calorimetry in food research – a review. *Food Chemistry*, **10**, 239–265

Burgess, G.D. and Stanley, D.W. (1976) A possible mechanism for thermal texturization of soybean protein. *Canadian Institute of Food Science and Technology*, **9**, 228–231

Camire, M.E., Camire, A. and Krumhar, K. (1990) Chemical and nutritional changes in foods during extrusion. *Critical Reviews in Food Science and Nutrition*, **29**, 35–57

Cheftel, J.C. (1986) Nutritional effects of extrusion-cooking. *Food Chemistry*, **20**, 263–283.

Dale, N.M., Araba, M. and Whittle, E. (1987) Protein solubility as an indicator of optimum processing of soybean meal. In *Proceedings Nutrition Conference for the Feed Industry, Georgia*, pp. 88–95

Eriksson, C.E. (1982) Lipid oxidation catalysts and inhibitors in raw materials and processed foods. *Food Chemistry*, **9**, 3–19

Falchowsky, G. and Schubert, R. (1993) *Vitamine unde weitere Zusatstoffe bei Mensch und Tier, 4. Symposium Jena/Thüringen*, Wissenschaftlicher Fachverlag Dr. Fleck, Niederkleen, Germany

Franks, F. (1988) Description and classifcation of proteins. In *Characterization of Proteins*, Edited by F. Franks. Humana Press, Clifton, New Jersey

Hager, D.F. (1984) Effects of extrusion upon soy concentrate solubility. *Journal of Agricultural and Food Chemistry*, **32**, 293–296

Hendriks, W.H., Moughan, P.J., Boer, H. and Poel, van der A.F.B. (1994) Effects of extrusion on the dye-binding, fluorodinitrobenzene-reactive and total lysine content of soyabean meal and peas. *Animal Feed Science and Technology*, **48**, 99–109

Hodge, J.E. (1953) Dehydrated foods: chemistry of browning reactions in model systems. *Journal of Agricultural and Food Chemistry*, **1**, 928–943

Hsu, H.W., Vavak, D.L., Satterlee, L.D. and Miller, G.A. (1977) A multi-enzyme technique for estimating protein digestibility. *Journal of Food Science*, **42**, 1269–1273

Hurrell, R.F., P.A. Finot and J.E. Ford. (1983) Storage of milk powder under adverse conditions. *British Journal of Nutrition*, **49**, 343–364

Hurrell, R.F. (1990) Influence of the Maillard reaction on the nutritive value of foods. In *The Maillard reaction in Food Processing, Human Nutrition and Physiology*, pp. 245–258. Edited by P.A. Finot, H.E. Aeschbacher, L.F. Hurrell and R. Liarton. Birkhauser Verlag, Basel

Jaenicke, R. (1965) Wärmaggregation und Wärmdenaturierung von Proteinen. In *Wärmebehandlung von Lebensmitteln. Symposions, Frankfurt (Main), 31 Mærz – 2 April*, pp. 207–244

Jager, T., Zuilichem, van D.J., Stolp, W. and van't Riet, K. (1989) Residence time distribution in extrusion cooking, Part 5: the compression zone of a conical, counter-rotating, twin-screw extruder processing maize grits. *Journal of Food Engineering*, **9**, 203–218

Jurgens, A., van Maanen, F.F.C., Vooijs, A.J. and Beumer, H. (1993). Differential Scanning Calorimetry (DSC) as a screening method for treated and untreated legume seeds and for pre-screening and monitoring methods for eliminating ANFs. In *Recent advances of research in antinutritional factors in legumes seeds*, pp. 477–480. Edited by A.F.B. van der Poel, J. Huisman, and H.S. Saini. Wageningen Press, Wageningen

Kato, H. (1987) Nutritional and physiological effects of Maillard reaction products. In *Trends in Food Science Proceedings of the 7th World Congress of Food Science and Technology, Singapore*, pp. 3–7

Kitabatake, N. and Doi, E. (1992) Denaturation and texturization of food protein by extrusion cooking. In *Food Extrusion Science and Technology*, pp. 361–371. Marcel Dekker Inc., New York

Labuza, T.P. (1971) Kinetics of lipid oxidation in foods. *Critical Reviews in Food Technology*, **2**, 355–405

Liener, I.E. (1994) Implications of antinutritional components in soybean foods. *Critical Reviews in Food Science and Nutrition*, **37**, 34–67

Marks, J. (1975) *A guide to the vitamins. Their role in health and disease.* Medical and Technical Publishing Co. Ltd, Lancaster, United Kingdom

Marsman, G.J.P., Gruppen, H. and Poel, van der A.F.B. (1993) Effect of extrusion on the *in vitro* digestibility of toasted and untoasted soybean meal. In *Recent advances of research in antinutritional factors in legumes seeds*, pp. 461–465. Edited by A.F.B. van der Poel, J. Huisman, and H.S. Saini. Wageningen Press, Wageningen

Marsman, G.J.P., Gruppen, H., Zuilichem, van D.J., Resink, J.W. and Voragen, A.G.J. (1995) The influence of screw-configuration on the *in vitro* digestibility and protein solubility of soybean and rapeseed meals. *Journal of Food Engineering*, **26**, 13–28

Nawar, W.W. (1985) Lipids. In *Food Chemistry*, pp. 139–244. Edited by O.R. Fennema. Marcel Dekker Inc., New York

O'Brien, J.M. and Labuza, T.P. (1994) Symposium provides new insights into nonenzymatic browning reactions. *Food Technology*, **July**, 56–58

Pederson, B. and Eggum, B.O. (1983) Prediction of protein digestibility by an *in vitro* enzymatic pH–STAT procedure. *Zeitung für Tierphysiology, Tierernährung und Futtermittelkunde*, **49**, 265–277

Poel, van der A.F.B., Stolp, W. and Zuilichem, van D.J. (1992) Twin-screw extrusion of two pea varieties: Effects of temperature and moisture content level on antinutritional factors and protein dispersibility. *Journal of Science and Food Agriculture*, **58**, 83–87

Poel, van der A.F.B., Huisman, J. and Saini, H.S. (1993) *Recent advances of research in antinutritional factors in legumes seeds*, Wageningen Press, Wageningen, The Netherlands

Schai, E. and Gadient, M. (1994) Effekt hydrothemischer Verfahren auf Futterzusatzstoffe. *Kraftfutter*, **5**, 162–164

Sheard, P.R., Ledward, D.A. and Mitchell, J.R. (1984) Role of carbohydrates in soya extrusion. *Journal of Food Technology*, **19**, 475–483

Stanley, D.W. (1989) Protein reaction during extrusion processing. In *Extrusion cooking*, pp. 321–341. Edited by C. Mercier, P. Linko and J.M. Harper, American Association of Cereal Chemists Inc., St. Paul, USA

Van Soest, P.J. (1976) Development of a comprehensive system of feed analysis and its application to forages. *Journal of Animal Sciences*, **26**, 119–128

Van Zuilichem, D.J., Jager, T. and Stolp, W. (1988). Residence time distributions in extrusion cooking. Part II: Single-screw extruders processing maize and soya. *Journal of Food Engineering*, **7**, 197–210

Visser, A., and Tolman, G.H. (1993). The influence of various processing conditions on the level of antinutritional factors in soya protein products and their nutritional value for young calves. In *Recent advances of research in antinutritional factors in legumes seeds*, pp. 447–553. Edited by A.F.B. van der Poel, J. Huisman, and H.S. Saini. Wageningen Press, Wageningen

Waldroup, P.W. and Smith, K.J. (1989) Animal feed uses of legumes. In *Legumes, chemistry, technology, and human nutrtion*, pp. 245–255. Edited by R.H. Matthews. Marcel Dekker, Inc., New York

6

LEGISLATION AND ITS EFFECT ON THE FEED COMPOUNDER

B.G. VERNON[1], J. NELSON[2] AND E.J. ROSS[1]
[1] BOCM Pauls, P.O. Box 39, 47 Key Street, Ipswich, UK
[2] UKASTA, 3 Whitehall Court, London, UK

New legislation and amendments to current legislation are discussed under the headings given in the text. The topics presented cover a broad selection of the legislation set in place in 1994 as well as those issues that are currently under discussion with the relevant legislative bodies.

Feeding Stuffs (Amendment) Regulations 1993

The labelling provisions of these amending regulations were brought into operation on 30 June 1993. Schedule 2 introduced the 'non-exclusive list of the principle ingredients normally used in compound feedingstuffs for all animals other than pets'. This list has to be used when individual ingredients, as opposed to categories, are declared. In particular the compounder must use the names given in the list on the statutory statement/feed label when incorporating any ingredients on the list.

Since the names in some cases were long, UKASTA developed a list of abbreviated terms for a number of these materials. LACOTS have advised that the use of these agreed abbreviated terms is acceptable. The use of non-agreed abbreviated terms might lead LACOTS to reconsider their use. The use of terms such as UK wheat must be labelled as just wheat and High Grade soya must be defined on the label as the abbreviated term soya or soya(bean) extracted, toasted; the latter being the full name for soya 48/49 extractions.

A list of the agreed abbreviated names is given in Table 6.1.

In this amendment were also additional controls on highly contaminated consignments of raw materials. In particular the level of aflatoxin B_1 in the named six raw materials and the level of arsenic and cadmium in phosphate

Table 6.1   LIST OF ABBREVIATED TERMS USED IN SCHEDULE 2 OF THE
FEEDING STUFFS (AMENDMENT) REGULATIONS 1993

Name	*Abbreviated Terms*
Oat Middlings	Oatfeed
Oat hulls and brans	
Barley middlings	Barley feed
Rice bran (brown)	
Rice bran (white)	Rice bran
Rice bran with calcium carbonate	
Rye middlings	Rye feed
Rye feed	
Wheat middlings	Wheat feed
Wheat feed	
Wheat bran	Bran
Maize germ expeller	Maize germ
Maize germ extracted	
Maize gluten feed	Maize gluten
Maize gluten	60% Maize gluten
Groundnut, partially decorticated, expeller	Ground nut exp
Groundnut, decorticated, expeller	
Groundnut, partially decorticated, extracted	Ground nut ext
Groundnut, decorticated, extracted	
Rape-seed, expeller	Rape meal
Rape-seed, extracted	
Safflowerseed partially decorticated, extracted	Safflower ext
Copra, expeller	Copra exp
Copra, extracted	Copra ext
Palm kernel expeller	Palm kernel exp
Palm kernel, extracted	Palm kernel ext
Soya (bean) heat treated	Full fat soya
Soya (bean) extracted, toasted	Soya
Soya (bean) dehulled, extracted, toasted	Hipro soya
Soya (bean) oil	Soya oil
Soya (bean) hulls	Soya hulls
Sunflower seed, extracted	Sunflower meal
Linseed expeller	Linseed meal
Linseed, extracted	
Pea middlings	Pea feed
Horse beans	Beans
(Sugar) Beet molasses	Molasses
(Sugar) Beet pulp, molassed	Beet pulp
(Beet) Sugar	Sugar
Manioc	Manioc or Tapioca
Carob pods	Carob pods or Locust bean
Lucerne meal	Lucerne
Wheat straw, treated	NIS
(Sugar) Cane molasses	Molasses
(Cane) Sugar	Sugar
Milk Skimmed Powder	Skimmed milk powder
Calcium carbonate	Chalk or Limestone
Dicalcium phosphate	Dical
Sodium chloride	Salt

sources were addressed. The maximum permitted content of arsenic and cadmium in phosphates is set at 20 mg/kg and 10 mg/kg respectively. The maximum level of aflatoxin B_1 in the named six raw materials has not changed but there are new requirements as to the handling of these materials.

The amendment prohibits the mixing of any ingredient(s) with other feedingstuffs if they contain specific undesirable substances above the permitted level as defined above as well as in Part II of Schedule 5 of the Feeding Stuffs Regulations (1991).

In addition the amendment also states that:

'No person shall import into Great Britain, sell or otherwise supply, or have in possession with a view to selling or otherwise supplying, any ingredient, unless that ingredient is sound, genuine and of merchantable quality'.

An ingredient would not be sound, genuine and of merchantable quality if the level of any substances contained in the said ingredient was such that, if incorporated into a compound feedingstuff, the level of any undesirable substance it might contain exceeded the maximum permitted limit for compound feeding stuffs.

In addition, the amendment states that:

'Any person engaged in the manufacture, sale, supply, importation or use of feedingstuffs or ingredients; or, when as a result of professional activities, a person possesses, has possessed, has had direct contact with, or is aware that an ingredient does not comply with the above requirements, shall immediately inform the authorities'.

For example, if any person knows about a consignment of one of the named six raw materials with an aflatoxin B_1 level above 0.20 mg/kg they must inform the authorities otherwise they could equally be prosecuted.

Feeding Stuffs (Amendment) Regulations 1994

In this amendment to the Feeding Stuffs Regulations (1991) an amendment was made to Schedule 7 with regard to the use and declaration of ammonium salts and analogues of amino acids.

In the case of ammonium sulphate (min 35%) use in ruminants, from the start of rumination, there are specific declaration requirements to be made on the label or packaging of the product:

- the words 'ammonium sulphate'
- nitrogen and moisture content
- animal species
- in case of young ruminants, the incorporation rate in the daily ration may not exceed 0.5%

In the case of the label or packaging of the compound feedingstuff:

- the words 'ammonium sulphate'
- the amount of the product contained in the feedingstuff
- percentage of the total nitrogen provided by non-protein nitrogen
- indication in the instructions for the use of the level of total non-protein nitrogen which should not be exceeded in the daily ration of each animal species
- in case of young ruminants, the incorporation rate in the daily ration may not exceed 0.5%

In relation to DL-methionine (min 65% copolymer vinylpyridine/styrene: max 3%) use in dairy cows the following declaration is to be made on the label or packaging of the product:

- protected methionine with copolymer vinylpyridine/styrene
- DL-methionine and moisture content
- animal species

In relation to L-lysine + DL-methionine (min 50%: including DL-methionine min 15% copolymer vinylpyridine/styrene: max 3%) use in dairy cows the following declaration is to be made on the label or packaging of the product:

- the name 'mixture of L-lysine monohydrochloride and DL-methionine protected with copolymer vinylpyridine/sytrene'
- L-lysine, DL-methionine and moisture contents
- animal species

In relation to the use of hydroxy analogues of methionine in all animal species the following declarations are required for the hydroxy analogue of methionine (min 85%: monomer acid min 65%) and the calcium salt of the hydroxy analogue of methionine (monomer acid min 83%: calcium min 12%):

On the label and packaging of the product:

- the appropriate name
- monomer acid and total acids content as given above for each product
- moisture content
- animal species

On the label or packaging of the compound feedingstuff:

- the appropriate name
- monomer acid and total acids content as given above for each product
- amount of the product contained within the feedingstuff

Feeding Stuffs (Amendment) (No. 2) Regulations 1994

This new legislation was brought into operation from 1st October 1994 for implementation on 1st January 1995. The legislation legitimised the current sale, supply and use of enzymes and micro-organisms (defined as additives in the Feeding Stuffs Regulations). This amendment also updates a number of regulations and schedules in the Feeding Stuffs Regulations (1991).

As part of the authorisation procedure for enzymes and micro-organisms, manufacturers had to submit to the authorising bodies of their individual EU Member States identification notes on the individual additives by 15th October 1994. The European Commission has undertaken to publish a list of enzyme preparations and micro-organisms permitted for use within the EU by 31st March 1996 and a full authorisation by 1st January 1997. In the UK the list of products for which identification notes have been accepted has not been published at the time of writing. In view of this LACOTS have agreed to a six months' transition period (from 1st January 1995) on the labelling requirements when incorporating authorised enzymes or micro-organisms into a feedingstuff. However, it should be noted that a company must be able to demonstrate that they are taking steps to comply with the legislation, since no person shall sell, or have in possession with a view to sale for use as a feedingstuff, or use as a feedingstuff, any material containing any enzyme or micro-organism, or sell, or have in possession with a view to sale for incorporation in a feedingstuff, unless these additives are authorised.

There are specific declaration requirements as follows:

ADDITIVE:

Enzyme:

- the names of the active constituents according to their enzymatic activities
- the identification number allotted by the International Union of Biochemistry (IUB)

- the name or business name and address or registered place of the business of the person responsible for the particulars referred to for the product
- the name or business name and address or registered place of the business of the person if he is not responsible for the particulars in the label or mark
- the activity units (expressed as activity per gramme or activity per millilitre [by an official or scientifically valid method])
- an indication of the period during which the activity will remain present
- the batch reference number and date of manufacture
- directions for use, indicating any safety considerations
- the net weight of any non-liquid additive
- either the net weight or the net volume of any liquid additive
- an indication of any significant characteristics listed in Part X of the Table in Schedule 4

Micro-Organisms

- the identification of the strain(s) according to a recognised international code of nomenclature
- the deposit number of the strain(s)
- the number of colony forming units (expressed as CFU/g)
- the name or business name and address or registered place of the business of the person responsible for the particulars referred to for the product
- the name or business name and address or registered place of the business of the person if he is not responsible for the particulars in the label or mark
- an indication of the period during which the colony forming units will remain present
- the batch reference number and date of manufacture
- directions for use, including any safety recommendations
- the net weight of any non-liquid additive
- either the net weight or the net volume of any liquid additive
- an indication of any significant characteristics arising during manufacture

PREMIXTURES:

Enzymes

- the names of the active constituents according to their enzymatic activities
- the identification number allotted by the IUB
- the name or business name and address or registered place of the business of the person if he is not responsible for the particulars in the label or mark
- the activity units (expressed as activity per gramme or activity per millilitre [by an official or scientifically valid method])
- an indication of the period during which the activity will remain present
- an indication of any significant characteristics listed in Part X of the Table in Schedule 4

Micro-Organisms

- the identification of the strain(s) according to a recognised international code of nomenclature
- the deposit number of the strain(s)
- the number of colony forming units (expressed as CFU/g)
- the name or business name and address or registered place of the business of the manufacturers if he is not responsible for the particulars in the label or mark
- an indication of the period during which the colony forming units will remain present
- an indication of any significant characteristics arising during manufacture

COMPOUND FEEDINGSTUFF:

The following declaration will have to be made in the Statutory section of the label.

Enzyme

- the names of the active constituents according to their enzymatic activities

- the identification number allotted by the IUB
- the activity units (expressed as activity per kilogramme or activity per litre [by an official or scientifically valid method])
- an indication of the period during which the activity will remain present
- an indication of any significant characteristics listed in Part X of the Table in Schedule 4

Micro-Organisms

- the identification of the strain(s) according to a recognised international code of nomenclature
- the deposit number of the strain(s)
- the number of colony forming units (expressed as CFU/kilogramme) if the number is measurable by an official or scientifically valid method)
- an indication of the period during which the colony forming units will remain present
- an indication of any significant characteristics arising during manufacture

Pesticides (Maximum Residue Levels In Crops, Food and Feeding Stuffs) Regulations 1994

An amendment to the Undesirable Substances and Products in Animal Nutrition Directive introduced maximum permitted levels for a range of pesticides in feedingstuffs and these were set out in the Feeding Stuffs Regulations (1991). Further control on the maximum pesticide residue levels were introduced in this new regulation.

This regulation sets out the maximum pesticide residue level for a wide range of products listed below:

Citrus fruit
Tree nuts
Pome fruits
Stone fruits
Vegetables
Pulses

Oilseeds
Cereals
Products of Animal Origin

This legislation is being updated on a regular basis as further information is made available within the EU.

A survey of pesticides began in 1990 as part of the 'grain passport' system at BOCM PAULS mill intake. The passport system involves the document-ation which accompanies the cereal and must list what, if any, pesticide has been used to treat the grain. The testing of pesticides in pig/poultry finished products is also carried out as a consequence of the large inclusion of cereals in such diets as part of the 'due diligence' approach under the Food Safety Act (1990).

The commonest pesticide found in cereals is pirimiphos-methyl. The next commonest pesticides found are alpha, beta and gamma hexachloro-cyclohexane (HCH) (see Table 6.2).

In the case of the 1159 parts per billion (ppb) pirimiphos-methyl level in wheat this was followed up since it has been the highest level found to date. An audit of the passport system identified the store from which the grain was delivered. It was ascertained that the cereal had been treated in the store as a result of insect infestation. It should be noted that the maximum residue level for pirimiphos-methyl is 5000 ppb. The maximum HCH

Table 6.2   RESULTS OF BOCM PAULS PESTICIDE ANALYSIS (EXPRESSED AS PARTS PER BILLION (PPB)

Pesticide	Cereal*	Mean	Std	Min	Max	No
NON TREATED	WHEAT	0	0	0	0	29
	BARLEY	0	0	0	0	14
TREATED						
PIRIMIPHOS-	WHEAT	112.6	213.6	6	1159.0	32
METHYL	BARLEY	29.4	50.86	5.0	207.0	14
ALPHA-HCH	WHEAT	1.0	0	1.0	1.0	1
BETA-HCH	WHEAT	3.0	0	3.0	3.0	1
GAMMA-HCH	WHEAT	6.7	4.11	2.0	12.0	3
Pesticide	Feedingstuff	Mean	Std	Min	Max	No
PIRIMIPHOS	PIG	122.6	92.49	0	301.0	27
METHYL	POULTRY	89.6	52.00	44.0	185.0	5

*Cereal tested as part of the grain passport procedure.
Legislation maximum residue limits (MRL): Cereals
Pirimiphos methyl 5000ppb, a + b HCH 20ppb and g HCH 100ppb

pesticide level defined in the Regulations is 20 ppb combined for alpha+beta HCH and 100 ppb for gamma HCH. The level and occurrence of other pesticides in animal feedingstuffs have been presented (Vernon and Ross, 1994).

Feeding Stuffs (Sampling and Analysis)(Amendment) Regulations 1994

This regulation was brought into operation on 11th July 1994 and introduced a new official European method for the determination of fibre that is different from the existing technique described in the Feeding Stuffs (Sampling and Analysis) Regulations 1982.

A collaborative trial was carried out by members of the UKASTA Analysts Committee to determine the extent of differences in the results between the two techniques. In the trial, eleven raw materials and seven finished products were analysed. Some of the results are presented in Table 6.3.

There was a good agreement between the two techniques for both raw materials and finished products except for rapeseed extractions and palm kernel. The differences in the fibre content of rapeseed extractions and palm kernel should be taken into account when formulating diets, in particular ruminant diets, and setting the fibre declaration. If a ruminant diet contains say 20% rapeseed and 10% palm kernel then, using the new figures presented in Table 6.3 the fibre declaration will be 0.82% units higher compared with the existing UK technique.

Table 6.3   THE FEEDING STUFFS (SAMPLING AND ANALYSIS) (AMENDMENT) REGULATIONS 1994—COMPARISON OF NEW METHOD FOR THE DETERMINATION OF FIBRE WITH THE CURRENT UK METHOD

	Fibre (%)	
	Current UK method	*New EC Method*
Raw Material		
Wheat	2.3	2.3
Rape Ext	10.0	11.8
Palm Kernel	16.0	20.6
Maize Gluten	7.5	7.2
Finished Product		
Broiler	3.0	2.9
Dairy	11.6	11.8
Sow	6.9	6.8

Bovine Spongiform Encephalopathy (BSE)

Throughout 1994 there has been continuing investigation into the origin of BSE as well as cooperation between the Ministry of Agriculture, Fisheries and Food (MAFF), UKASTA, feed compounders and the rendering industry on the various new pieces of legislation and amendments to existing legislation related to BSE. In this section a number of pieces of legislation are discussed.

AMENDMENTS TO THE BOVINE OFFAL (PROHIBITION) REGULATION 1989

The Spongiform Encephalopathy Advisory Committee (SEAC) reviewed the use of tallow derived from Specified Bovine Offal (SBO) as well as such tallow arising as a by-product from the rigorous processing in the oleochemical industry. SEAC concluded that material obtained from SBO-derived tallow, as a consequence of processing by the oleochemical industry could continue to be used in animal feedingstuffs but not other SBO derived tallow. MAFF propose to consolidate all legislation on SBO shortly.

AMENDMENTS TO THE BOVINE SPONGIFORM ENCEPHALOPATHY ORDER 1991

As part of the continuing review of BSE, the possible effect of the disease in calves has been studied. As a result of this study it was stated by SEAC that the potential risk to human health from food from infected calves was minuscule. The UK Government's policy of caution over BSE, however, resulted in an extension of the existing SBO regulations to ban the use of the intestine and thymus from calves under six months of age. This action is considered precautionary and will be kept under review.

Following a review within the EU these amendments, which apply to all Member States, extend the original UK ban on the feeding of ruminant meat and bone to ruminants (Bovine Encephalopathy Order, 1988) to banning the feeding of non-ruminant meat and bone to ruminant animals. The term non-ruminant in this case applies to meat and bone of ruminant (bovine and ovine) and porcine origin; i.e. protein derived from mammalian tissues.

SAMPLING AND TESTING OF RUMINANT FEEDINGSTUFFS FOR
THE PRESENCE OF RUMINANT PROTEIN

As part of the continuing investigation into BSE the MAFF have developed
an enzyme-linked immunosorbent assay (ELISA) test to detect the presence
of ruminant proteins in rendered animal material (Ansfield, 1994). The
technique can differentiate between bovine/ovine and porcine meat and
bone. This test will be used as part of a voluntary sampling and analysis
procedure when carrying out routine examinations of BSE-suspect cattle.
    The limit of detection of this ELISA test is as follows:

1 part of ruminant meat and bone in 400 parts of a feedingstuff (ie.
0.25%)
1 part of ruminant meat and bone in 6000 parts of meat and bone (ie.
0.0167%)

APPROVAL OF ALTERNATIVE HEAT TREATMENT SYSTEMS FOR
PROCESSING ANIMAL WASTE OF RUMINANT ORIGIN, WITH A
VIEW TO THE INACTIVATION OF SPONGIFORM
ENCEPHALOPATHY AGENTS (COMMISSION DECISION) 1994

This EU decision (EC 1994) for all Member States was made in June 1994
for implementation from 1st January 1995. It sets out different time/tem-
perature profiles and other parameters for the rendering of ruminant meat
and bone. An example of one of the time/temperature requirements is
given in Table 6.4.
    The different plants, whether these are batch or continuous processes,
will be validated within the UK by the MAFF as per one of the specified

Table 6.4   EXAMPLE OF TIME/TEMPERATURE REQUIREMENTS FOR
RENDERING OF RUMINANT MEAT AND BONE

Chapter V/VI: Continuous/Atmospheric/Added Fat and Continuous/Pressure/Added
Fat 30 mm particle size maximum

Temperature (°C)	>100	>110	>120	>130
Time (mins)	16	13	8	3

processing conditions. As part of the validation process a procedure for the identification of hazard and critical control points (HACCP protocol) will have to be carried out. Where plants meet the validation criteria a certificate will be issued by MAFF.

Dietetic Feedingstuffs Directive

This directive was adopted in July 1994 and establishes the list of intended uses of a dietetic feedingstuff. A dietetic feedingstuff is a feed which has a particular nutritional purpose. The definition of a 'particular nutritional purpose' means the purpose of satisfying the specific nutritional needs of certain pets and productive livestock whose process of assimilation, absorption or metabolism could be temporarily impaired or is temporarily or irreversibly impaired and therefore able to derive benefit from the ingestion of feedingstuffs appropriate to their condition.

The list of authorised dietetic feedingstuffs includes the following as examples:

Reduction of the risk of milk fever
Reduction of the risk of ketosis
Reduction of the risk of tetany (hypomagnesaemia)
Reduction of the risk of acidosis
Stabilisation of water and electrolyte balance
Reduction of the risk of urinary calculi
Reduction of stress reactions
Stabilisation of physiological digestion
Reduction of the risk of constipation
Reduction of the risk of fatty liver syndrome
Compensation for malabsorption
Compensation for chronic digestive disorders of small intestine
Reduction of the risk of digestive disorders of large intestine
Compensation of electrolyte loss in cases of heavy sweating

There are compulsory labelling requirements laid down in the Directive. These are in addition to the provisions already laid down for compound feedingstuffs and are as follows:

- the qualifying expression 'dietetic' together with the description of the feedingstuff
- the precise use, i.e. the particular nutritional purpose
- the indication of the essential nutritional characteristics of the feedingstuff

- the declaration of specific ingredients
- the recommended length of time for use of the feedingstuff

An example of the declaration requirements for a dietetic feedingstuff to reduce the risk of urinary calculi is given in Table 6.5.

Table 6.5   EXAMPLE OF THE DECLARATION REQUIREMENTS FOR A DIETETIC FEEDSTUFF TO REDUCE THE RISK OF URINARY CALCULI

— Nutritional Purpose
  — Reduction of urinary calculi
— Essential Characteristics
  — Low phos/mag and urine acidifiers
— Species
  — Ruminants
— Labelling (Total = Added Background)
  — Cal/Phos/Sod/Mag/Potassium/Chlorine/Sulphur and urine acidifiers
— Time
  — Up to 6 weeks
— Other Provisions

Proposal for a Council Regulation Laying Down the Conditions for Approving Certain Establishments in the Animal Feed Sector

This regulation is at the discussion stage and lays down the conditions and arrangements for approving certain categories of establishments in the animal feed sector. The registration refers to establishments manufacturing additives, premixtures and/or compound feedingstuffs containing pre-mixes or additives. In addition, it would cover home-farm mixers who circulate feed to third parties or use premixtures or additives, as would those using raw materials containing certain undesirable substances. Intermedi-aries in the distribution chain would also be required to register. Any company that operates on several sites would have to register each estab-lishment individually.

The proposal lays down criteria for a site's facilities, personnel, produc-tion, quality control, storage/distribution, documentation and com-plaints/product recall. For each site a registration fee will have to be paid and each establishment will be given a unique identification number which will have to be printed on the label of the manufactured product. This proposal for registration of establishments is a topic to be found in a number of proposed EU legislations for the production of animal feedingstuffs that are discussed below.

Proposal Concerning the Authorization and Circulation of Specific Feed Materials

This proposed directive is an amendment to the Certain Products Directive (1982) and its amendments which relates to products which have a nitrogenous character and are put into circulation as a feedingstuff or in a feedingstuff.

In the new proposal there are four categories of specific feed materials:

A: i. Materials and products subjected to decontamination or a detoxification process
    ii. Products resulting from genetic engineering

B: Physically/chemically treated products and by-products from primary agriculture, the agricultural processing industry and the processing food/non-food industry

C: Chemical Process

D: Biomass products and by-products from microbial activity

There is reference to scientific and technological developments in the food and non-food industry as well as in the agricultural processing industry. In addition, there is a requirement to examine the safety risk for humans, animals and the environment. The effect of these criteria for each of the above categories could result in raw materials already in use, with no known problems, having to suddenly fulfil the requirements of this proposed directive.

For example, under category Ai full fat soya could be classified under the detoxification process, i.e. destruction of the trypsin inhibitor. In category B, materials subjected to long standing physical and chemical treatment such as sodium hydroxide treated straw do not need to be covered by this proposal.

It has been proposed to the European Commission that only new materials coming forward should be subject to this directive. For example, the producers of genetically modified plants should provide a safety dossier for their use in both human and animal feedingstuffs. In addition, if a material has already been given clearance for use in human food it should not be subject to further clearance for the possible use in animal feeds.

Proposal on the Circulation of Feed Materials

This proposed directive is designed, in principle, to harmonise the Straights Feeding Stuffs Directive (1977). This new directive is designed to regulate the marketing of both straight feedingstuffs and raw materials used in the manufacture of animal feedingstuffs. The declaration of a variety of constituents is also required by the seller of the material at the beginning of the feed chain rather than carrying out unnecessary multiple analyses just before the end of the feed chain, whether this is carried out by a compounder, home-mixer or farmer.

An important feature of this proposal is the requirement for each consignment of a feed material, whether delivered in bag or bulk, to be accompanied by a label containing specific information as listed in part B and C of the Directive. Part B refers to a non-exclusive list of the main feed materials and part C describes provisions for the declaration of certain constituents not listed in part B. Part B describes individual raw materials split into one of twelve categories (see Feeding Stuffs (Amendment) Regulations (1993) for the categories), the name of the material and the compulsory declaration items (Table 6.6).

In part C, however, there is a category listing with the required declaration items (Table 6.7).

The requirement to label each consignment is the same as the provision in operation for delivery of compound feedingstuffs but differs from the controls currently in place for straight feedingstuffs. This is also different from the Feeding Stuffs Regulations (1991) which permit, for straight feedingstuffs delivered in bulk, the statutory statement containing the labelling information to be given as soon as practicable after delivery to the purchaser.

The accompanying label will have to include the following declaration:

- the word 'feed material'
- the name of the feed material
- the specific declared constituents
- moisture content if above 14.5%
- acid insoluble ash if above 2.2% in the dry matter
- net quantity
- the name and address of the person responsible for the declaration

Provisions have also been made in the directive for the minimum tolerated levels for the analytical constituents which are declared as required under part B or part C.

There is the provision for a different declaration on the label dependent

Table 6.6   EXAMPLE OF RAW MATERIAL DESCRIPTION FROM PART B OF THE
PROPOSED CIRCULATION OF FEED MATERIALS DIRECTIVE

1: Cereal Grains, their Products and By-Products			
Number	Name	Description	Compulsory Declaration
1.08	Rice Bran (brown)	By-product of the first polishing of dehusked rice. It consists principally of silvery skins, particles of the aleurone layer, endosperm and germ.	Crude fibre Crude fat

Table 6.7   EXAMPLE OF RAW MATERIAL DESCRIPTION FROM PART B OF THE
PROPOSED CIRCULATION OF FEED MATERIALS DIRECTIVE

Feed Material Belonging to:	Compulsory Declaration of:
Products and by-products of tubers, roots	Starch, Crude Fibre
Forage and roughage	Crude Protein, crude fibre
Products and by-products of the sugar cane processing industry	Crude protein, Crude fibre, total sugar expressed as sucrose

on whether the recipient of the material is a registered feed compounder or
not. For example,

NON-REGISTERED COMPOUNDER

Label for Rice Bran (brown) should include:

- the word 'feed material'
- Rice Bran (brown)
- crude fibre and crude fat level
- moisture content if above 14.5%
- acid insoluble ash if above 2.2% in the dry matter
- net quantity
- the name and address of the person responsible for the declaration

REGISTERED COMPOUNDER

Label for Rice Bran (brown) should include:

- the word 'feed material' plus the name and address of the destined compound feed manufacturer
- Rice Bran (brown)
- net quantity
- the name and address of the person responsible for the declaration

The analytical constituents should be agreed between the seller and the buyer and defined in the purchase contract between the two.

Proposed Directive Fixing the Principles Governing the Organisation of Inspections in the Field of Animal Nutrition (Feeds Control)

This proposed Directive is designed to fix harmonised principles governing the organisation of inspectors in the field of animal nutrition. This would cover all stages from production, processing, storage, transport, distribution trade and use of feedingstuffs, as well as imports into the EU from third countries. The check function would also cover existing legislation such as the Directives on Additives, Undesirable Substances, Compounds, Straights, Dietetic Feedingstuffs and Certain Products as well as any future legislation.

The checks and inspections require

- documents to verify the product
- identification checks to confirm the product documentation, including labelling
- physical check to include analytical testing
- annual report by each Member State as to the result of the inspections and what actions were taken when infringements were found

Within the proposed Directive there are guidelines on the notification to authorising bodies within the Community of any problems. Where problems are identified with a particular raw material there are various options which would apply to the subsequent handling and/or use of the material, i.e. destruction in the case of a highly contaminated consignment of a raw material.

Proposed Amendment to Article 12 of the EC Additives Directive

This proposed amendment to the EC Additives Directive is designed to harmonise Article 12 by removing the existing derogations and requiring all Member States to introduce national legislation for the minimum incorporation of additives (non-medicinal and Permitted Merchant List (PML) drugs) when incorporated into complementary feedingstuffs. As with the current provisions, and as defined in this new amendment the incorporation of a higher level of additive than is established for complete feedingstuffs would be allowed provided that, when the complementary feedingstuff is diluted for feeding to livestock, the additive level does not exceed the maximum level fixed for the complete feedingstuff. The non-medicinal additive level in the UK is set out in Schedule 4 of the Feeding Stuffs Regulations (1991) and its amendments and for PML products as laid down in their individual product licences.

The proposed amendment sets out that the quantity of the complementary feedingstuff when fed or mixed should represent at least

- 2% of the daily ration for all species or category of animals excluding ruminants and equines
- 2% of the overall complementary feedingstuff in the case of ruminants and equines provided that the content of antibiotics do not exceed 2000 mg/kg

This proposal will have an effect on the use of such products within each Member State but it has important implications for the UK compound feed industry. In particular, the production of feed blocks and liquid feedingstuffs containing PML products may be prohibited. Further discussions are continuing on the implications of this proposal within each Member State.

Revised Code of Practice for the Control of Salmonella (1994)

Within the UK the MAFF has circulated a set of revised Codes of Practice for the Control of *Salmonella*.

- during the storage, handling and transport of raw materials intended for the incorporation into, or direct use as, animal feedingstuffs

- in the production of the final feed for livestock in premises producing less than 10,000 tonnes per annum
- in the production of the final feed for livestock in premises producing over 10,000 tonnes per annum

In essence they are similar to the existing codes of practice but there is now emphasis on the use of the principle of Hazard Analysis Critical Control Point (HACCP) investigations. The HACCP principles described in the code of practice are as follows:

- all areas and processes are assessed for 'risk'
- regular surveillance of certain Critical Control Points (CCP) are completed, with reference to non-compliance limits for each CCP. Each non-compliance is recorded and acted upon using a non-compliance system assigned to each CCP.
- records of all HACCP systems should be kept for observation by the authorities

The final revision of the new codes of practice are being prepared.

References

Ansfield. M (1994). Production of a sensitive immunoassay for detection of ruminant proteins in rendered animal material heated to >130°C. *Food & Agricultural Immunology*, **6**, 419–443

Bovine Spongiform Encephalopathy Order (1988). London: Her Majesty's Stationery Office

Bovine Spongiform Encephalopathy Order (1991). London: Her Majesty's Stationery Office

Bovine Offal (Prohibition) Regulation (1989). London: Her Majesty's Stationery Office

Certain Products Directive (1982). London: Her Majesty's Stationery Office.

EC (1994). Approval of Alternative Treatment Systems for Processing Animal Waste of Ruminant Origin with a View to the Inactivation of Spongiform Encephalopathy Agents (Commission Decision) (1994). EC Directive

The Feeding Stuffs Regulations (1991). London: Her Majesty's Stationery Office

The Feeding Stuffs (Amendment) Regulations (1993). London: Her Majesty's Stationery Office

The Feeding Stuffs (Amendment) Regulations (1994). London: Her Majesty's Stationery Office

The Feeding Stuffs (Amendment) (No:2) Regulations (1994). London: Her Majesty's Stationery Office

The Feeding Stuffs (Sampling and Analysis) Regulations (1982). London: Her Majesty's Stationery Office

The Feeding Stuffs (Sampling and Analysis)(Amendment) Regulations (1994). London: Her Majesty's Stationery Office

Food Safety Act (1990). London: Her Majesty's Stationery Office

Pesticides (Maximum Residue Level in Crops, Food and Feeding Stuffs) Regulations (1994). London: Her Majesty's Stationery Office

Straights Feeding Stuffs Directive (1977). London: Her Majesty's Stationery Office

Vernon, B.G. and Ross, E.J. (1994). The occurrence and control of toxins and undesirable substances in animal feeds. *45th EAAP Symposium, Edinburgh*, EAAP. In press.

III

Pet and Poultry Nutrition

7

SENSORY AND EXPERIENTIAL FACTORS IN THE DESIGN OF FOODS FOR DOMESTIC DOGS AND CATS

C.J. THORNE
WALTHAM Centre for Pet Nutrition, Freeby Lane, Waltham-on-the-Wolds, Melton Mowbray, Leicestershire. LE14 4RT, UK

The domestic cat (*Felis silvestris catus*) and the domestic dog (*Canis familiaris*) are members of the order Carnivora and have become an integral part of human society as evidenced by the large number of animal-owning households worldwide. In the UK alone it is estimated that there are in excess of 14 million domestic cats and dogs. One in every four households in Western Europe owns a dog and the figure rises to two in every five households in the USA. The adoption of pets in millions of homes has provided the impetus for the development of a sophisticated pet food industry with extensive research into the nutritional needs of their consumers (MacDonald, Rogers and Morris, 1984; National Research Council, 1985, 1986; Burger and Rivers, 1989). This has resulted in a pet food market that provides a great variety of commercially prepared, nutritionally complete foods which vary in their composition and format. The role of pet foods is somewhat different from other dietary regimes in that a single compounded food could be the sole provider of nutrients over many years and for all life-stages, compared with other long-term feeding regimes, such as our own, in which complete nutrition is usually obtained through consumption of a wide variety of food of differing nutrient composition. However, both dogs and cats show idiosyncrasies in their food preferences and understanding the mechanisms whereby they select between foods and the derivation of these food preferences are key areas of research. The following review will concentrate on the latter.

Chemical senses

Olfaction and taste are the primary senses in food selection with the mouth-feel (texture) of the food acting as a further factor in food selection.

151

The primary route for food selection is through the olfactory sense, with the odour of the food providing the animal with qualitative information as to the suitability of the dietary item. Flavour, the combined stimuli of odour and taste, of the dietary item provides further information on the quality and safety of the food with texture defining the handling characteristics.

OLFACTION

The dog's perception of its environment is very dependent on its acute olfactory sense. Forty years ago, Kalmus (1955) showed that the dog was able to discriminate identical twins if their scents were given together, but when given the scent of one twin would follow the scent-trail laid by the other. Hence the acuity of olfactory perception for the dog is similar to the human powers of visual perception. Anosmic dogs were able to distinguish meat from cereal, but their ability to distinguish between types of meat was much reduced, with their preference for the sweet taste of sugar undiminished (Houpt, Hintz and Shepherd, 1978). An appreciation of the importance of smell can be gained from the surface area of the olfactory epithelium, which is about $75\,cm^2$ in the beagle (Albone, 1984), with a range of 18 to $150\,cm^2$ across different breeds of dog, $21\,cm^2$ in cats, but only 3 to $4\,cm^2$ in man (Dodd and Squirrel, 1980). The potential for olfactory stimulation is far higher in both the dog and cat than in man, with the response in the cat being further enhanced through a very high density of neurones in the olfactory epithelium. Feeding behaviour can be significantly altered by odour. Cats maintained on a bland dry-food altered their pattern of food intake when a palatable meat aroma was presented such that food consumption was almost exclusive to those periods when the odour was present (Mugford, 1977a). However, the behavioural link between odour and food is not rigid, since meaty odours alone will not overcome neophobia in cats (Bradshaw, 1986) or sustain interest in bland food for dogs (Houpt, 1978). The sense of smell is used in two distinct ways, either on its own, as when an object is investigated by sniffing, or in conjunction with taste to provide full perception of food flavour. Dogs, and to a lesser extent cats, rely on their sense of smell to a much greater degree than humans, and yet it is the least well understood of all their senses.

TASTE

Although the sense of taste is less complex than that of smell, information on the chemical stimuli which evoke taste sensations is based primarily on studies of neurophysiology which are poorly supported by behavioural studies. Comparison of the taste abilities of the cat and the dog with those of other species provides differences which could indicate adaptations to a carnivorous lifestyle.

The most abundant taste units are those which are most sensitive to amino acids that are described by humans as having a 'sweet' taste. Cats prefer solutions of the 'sweet' amino acids which stimulate response in these units and reject the 'bitter' amino acids that inhibit the response (White and Boudreau, 1975; Beauchamp, Maller and Rogers, 1977). The same amino acids that are inhibitory in the cat tend to be either neutral or stimulating in the dog (Boudreau, Sivakumar, Do, White, Oravec and Hoang, 1985). These units are also sensitive to sodium chloride and potassium chloride and in the dog are also sensitive to sugars and respond to a wide range of monosaccharides and disaccharides and some artificial sweeteners. In the cat, neither this nor any of the taste systems have been found to respond to sugars at any behaviourally meaningful concentration (Boudreau, 1989).

A second group, the acid units, responds to carboxylic acids, phosphoric acids, and other Bronsted acids such as nucleotide triphosphates and are also stimulated by some amino acids such as the sulphur-containing L-cysteine and L-taurine (Boudreau *et al.*, 1985). Cat and dog units are generally similar, but cats reject medium-chain fatty acids (8:0) but not short-chain (MacDonald, Rogers and Morris, 1985), which may indicate an interaction between the acid units and other neural groups.

The remaining taste groups in the facial nerve are less well characterised, but in both cat and dog, all these units respond to nucleotide diphosphates and nucleotide triphosphates, but subgroups also respond to other compounds. In the cat, one subgroup is stimulated by a diverse range of substances, including quinine, and may be responsible for the rejection of quinine by cats at low concentration (Carpenter, 1956). In the dog, a subgroup responds to a narrow range of sweet-tasting substances, particularly furaneol and methyl maltol, suggesting at least two sweet taste systems as in man.

The preponderance of amino acid units in the dog and cat has been related to meat-eating, presumably to give them the ability to distinguish between meats of different quality. Monophosphate nucleotides, which accumulate after prey has been killed, are inhibitors of amino acid units in cats, and may regulate their feeding on carrion. The response of these units in the cat to taurine may reflect the cat's requirement for this amino acid, but the dog, which does not have this requirement, shows an even greater sensitivity to sulphur amino acids (Boudreau et al., 1985). A comparatively low sensitivity to NaCl in both species can be related to the nutritionally balanced sodium content of their natural prey. The omnivorous habits of the dog may have led to the retention of both the sweet tastes seen in man, of which the furaneol group has been speculatively connected with fruit-eating (Boudreau, 1989). It can be speculated that the strictly carnivorous cat has refined both its amino acid and nucleotide systems to exclude potentially confounding signals from compounds characteristic of plant foods.

FOOD TEXTURE

Although the flavour of a food is the primary driver of acceptance in both the dog and the cat, the form in which the food is offered is also important in selection. Dogs prefer a canned meat or semi-moist diet to a dry-food (Kitchell, 1978), canned-meat to the same meat freshly cooked and cooked meat to raw meat (Lohse, 1974). The role of texture is even more important in the cat, whose dentition is more specialised than that of the dog which is often used to represent the general carnivore pattern of dentition (Ewer, 1973). The dog, in common with most other canids, has a total of 42 teeth when adult. The incisors are fairly large and slightly curved, and are used for gripping and tearing flesh, along with the large canines. Of the pre-molars and molars, the last upper premolar and the first lower molar are most specialised. Known as the carnassials, parts of these teeth are laterally flattened, and act like shears as they are brought past each other. Behind the carnassials, the molars are adapted for crushing. In comparison, the adult cat has only 30 teeth, the reduction, compared with the dog, being entirely in the back of the jaw. This difference reflects the greater specialisation for flesh-eating in the cat with a total absence of flattened molars which are necessary for grinding vegetable materials. The incisors, in the cat, are small, and while they are used for a certain amount of ripping and scraping of meat, their primary function may be in grooming. The canines are long, sharp and laterally compressed, and their most specialised use is in dis-locating the neck vertebrae of prey. The cat is unable to chew effectively due to the reduction in dentition, specialisation of the individual teeth, and parallel shortening of the jaw, and reduces the size of its food by tearing or shearing into pieces that can be swallowed. The renowned 'finicky' feeding habits of the cat must, in part, be due to their specialisation as a flesh eater and the importance of the format of the food item in terms of ease of handling. For wild individuals the cost-benefit equation for procuring a meal would be very real and feral farm cats ate very little of a hard dry-food when a food with a high water content was available (Bradshaw, Macdonald, Healey and Arden-Clark, 1991), and this difference was ascribed to the longer 'handling' time required for the dry food.

Dietary selection

The life-style of the wild canids and felids is very different from that of the domestic dog and cat, and there is little evidence for an important role of palatability. The wild counterpart is not presented with a reliable supply of food each day, as is the domestic dog and cat, but has to expend considerable energy in locating and catching prey. In this situation the palatability of the prey is probably of lesser importance than knowing that the food is safe and meets the animal's nutritional needs. However, when food is readily available even wild carnivores will be selective. Although the route to obtaining food for the domestic cat and dog is very different from that of their wild ancestors, the underlying behavioural mechanisms on which food selection is based may still be intact, if modified somewhat by domestication. In-built patterns of behaviour appear to play a large part in discriminating the useful food items, as orphaned animals raised by hand, without the benefits of learning from their natural parents, will select food items appropriate to their life-style. The experiences provided by the natural parents appear to modify the food-selection behaviour of the young with the resultant development of a response more appropriate to their environment.

At least two alternative feeding strategies appear in infancy. On the one hand, there is the process of 'food imprinting' in birds (Hess, 1964) and 'the fixation of food habits' in both cats and dogs (Kuo, 1967), while on the other hand there is the tendency for laboratory rats (Morrison, 1974) and cats and dogs (Mugford, 1977a) to prefer foods having novel sensory characteristics. Both strategies would have adaptive significance, in that dietary restriction to known foods and avoiding unknown and potentially toxic items provides a safe strategy. Toxic edibles constitute one of the most serious challenges to survival of the young and this challenge may be met by preference for familiar over novel foods (ingestational neophobia) (Domjan, 1973). Alternatively, discovering and trying new sources of nutrients would provide a real advantage if the usual food source became scarce.

Food preferences could be the result of genetic predisposition which makes one flavour more palatable than another or they may develop as a result of environmental influences and experience. There is no unequivocal evidence for inherited food preference, but both animals and humans appear to have taste preferences that are directly related to the nutritional properties of the food. A sweet taste is often associated with a high concentration of carbohydrates and a bitter taste results from many toxic materials (LeMagnen, 1967). Almost all animals, with the exception of the cat, prefer the sweet taste over others (Soulairac, 1967) and many appear to have an innate preference for salt and an aversion to bitter-tasting foods (Denton, 1967; Rozin, 1967; Rozin and Kalat, 1971). Hence, some taste preferences are observed early in life, and appear to be relatively unaltered by subsequent learning experiences (Rozin and Kalat, 1971; Scott and Quint, 1946). Moreover, certain preferences can be selectively increased or decreased by inbreeding, thus suggesting that there is a genetic component to some food preferences and aversions (Scott, 1946).

On the other hand, experience strongly influences the selection of food, since odours, textures, volumes, and tastes often bear no clear relationship to nutritional value (LeMagnen, 1967; Rozin, 1967; Rozin and Kalat, 1971). For example, many animals rapidly decrease their food intake when a new, adequate diet is abruptly substituted for the familiar one; after a time, however, intake increases progressively and eventually becomes stabilised.

A variety of environmental influences are involved in the development of food preferences. Flavour experience, in both the young and adults, appears to be important and provides one route for the development of individual variability in food likes and dislikes. The feeding context is also important in that many young animals gain experience of safe foods by copying the adults. Associative learning in the form of a link between the flavour of the food and its nutritional consequences provides the means for identifying those foods which provide nutritional benefit, whilst avoiding those foods that are nutritionally detrimental. The need for this flexibility in food preferences is obvious from the wide range of prey items taken by feral cats (Fitzgerald, 1988), and the omnivorous habits of dogs. As with many other behaviours, the behavioural processes involved in food selection are modified in response to the experience gained throughout life.

Development of food preferences

The foetus is surrounded by amniotic fluid rich in chemicals of changing concentration and regularly swallows these fluids during development *in utero*. The neonate is suckled by the mother, whose milk will vary in quality and flavour dependent on her diet. In the wild, prior to weaning, pups will be offered regurgitated partly-digested food from the mother and kittens will be offered small pieces of prey providing their first experience of the local food supply. At weaning, the young will tend only to eat from those food sources used by their parents or other adults. Thus, there are many opportunities for the very young to experience and to link flavours with safe food sources and hence develop a food selection strategy based around those foods which are available, nutritious and safe. Food selection behaviour is a process which involves complex interaction of inherited or early-established food preferences with experience gained in local conditions of prey availability and capture. The home-life of the domestic cat and dog is very different from that of their wild ancestors, but similar developmental and experiential processes are involved in the development of food preferences.

THE MATURATION OF THE SENSE OF TASTE

It has been demonstrated by anatomical study of the tongue that the peripheral gustatory system of the dog was functional at birth, but had not yet fully developed (Ferrel, 1984a). This would indicate the possibility that the puppy may be responsive to chemical stimulation at or before birth. Ferrel (1984b) also investigated the responses of the gustatory nerves and found that *chorda tympani* nerve responses to chemical stimulation of the tongue were present at birth in the dog. It has been suggested that taste experience *in utero* is involved in the establishment of the taste preference and aversion behaviour expressed after birth (Minstretta, 1972; Minstretta and Bradley, 1977). Smotherman (1982) found that rats exposed *in utero* to apple solution showed an increased preference for that flavour. It seems reasonable to assume that similar experiential effects would occur in the dog and may be the first stage in the development of flavour preferences.

THE ROLE OF SUCKLING

During the nursing period the young are provided with food cues from the flavour of the milk, which in rat pups causes them to seek out and preferentially ingest the food that their mother had been eating (Galef and Henderson, 1972). The clinical literature indicates that a wide variety of substances, including antibiotics, sulfonamides and most alkaloids, when ingested pass intact into the nursing infant. Thus the complex long chain molecules associated with dietary taste and smell will also pass intact from the intestinal tract into her milk. Ling, Kan and Porter (1961) described changes in the flavour of cow's milk associated with the ingestion of certain natural foodstuffs. The potential for using food flavours which pass intact into the nursing infant has been exploited in pig rearing. A flavouring substance designed to be incorporated into the sow's feed and pass intact into the milk resulted in an increase in post-weaning food intake and growth rate for litters offered food with the same flavour added (Campbell, 1976).

EARLY EXPERIENCE OF SOLID FOOD

In many species, exposure to a specific dietary flavour early in life has been shown to enhance subsequent preference for that flavour. The strength and persistence of such a preference are influenced by such factors as the species studied, the animal's developmental age, the attractiveness of the flavour employed and the duration of exposure. For example, turtles and snakes exhibit long lasting preferences for the first fed food over subsequently presented foods, even if it is not a natural food prey item (Burghart and Hess, 1966; Fuchs and Burghart, 1971).

Several studies have investigated the role of early flavour experience on food choice in the dog and cat. Kuo (1967) hand-reared pups and kittens from birth to six months of age on extremely unusual diets. The pups fed on a soyabean diet would eat no novel food, those reared on a mixed vegetarian diet would eat no animal protein, but those reared on a mixed diet would eat any new food except those with a bitter, sour or stale taste. Limiting flavour experience in these pups led to a fixation of food habits, but provision of flavour and textural variety within the diet enhanced the acceptance of novel foods.

Our own studies, involving natural weaning and single food flavour experience, contradict Kuo's findings (Mugford, 1977a, 1977b). Basenji and Terrier pups fed a single canned food from weaning for the next 16 weeks preferred a novel food to their accustomed diet. The persistence of this preference was dependent on the relative palatability of the diets; novelty of the food combined with low relative palatability produced only a short-lived preference for the novel diet, whereas when the diets were of similar palatability the effect was more persistent.

A similar study was carried out by Ferrel (1984c) with three and a half week old Beagle pups offered single semi-moist foods with characteristic flavours for three weeks prior to preference testing. She noted considerable individual variability of response, but concluded that palatability and novelty were the important factors in diet selection, which is in agreement with our studies. However, one diet group showed a tendency to prefer the food with the accustomed flavour, but this was also confounded with a higher palatability for that diet.

It can be concluded from these data that if there is a sensitive period for the development of fixed flavour preferences it is before three and a half weeks of age. In these investigations the bitches were fed on a variety of diets and yet the pups responded to novel foods in a similar fashion suggesting little or no effect of the mother's milk on subsequent diet choice. The novelty and palatability of the food appear to be most important in controlling food choice highlighting the plasticity of food selection.

Flavour experience in the adult

In the adult cat and dog, flavour preferences continue to alter with new experience and several behavioural strategies for identifying adequate food sources can be observed in both domestic species. There appears to be little, if any, direct perception in cat or dog of the nutritional status of a food, but they will rapidly develop an association between a food's flavour and the physiological consequences of its consumption. Thus, learning occurs in a situation where the flavour experience and the physiological effect are well separated in time, a situation which is not easily reconciled with the usual necessity for a close temporal link between stimulus and reward in classical models of conditioning. This type of learning will result in a preference for flavours associated with nutritional benefit and an avoidance of flavours associated with ingestive negatives. Baker, Booth, Duggan and Gibson (1987) has demonstrated in rats that this flavour and nutrient association results in the development of nutrient specific hungers in which the preference is only observed when the rat is mildly deficient in that specific nutrient. If the rat is given a pre-load with the specific nutrient the preference for a flavour associated with that nutrient is not observed. Although preference for flavours associated with a particular nutrient has been demonstrated in both dog and cat, nutrient specific hungers have not.

NEOPHOBIA

Neophobia is a fear of new objects or a rejection of new foods. Food neophobia in dogs is not common, but has been demonstrated in our studies. A variety of small breeds, Poodles, Dachshunds, Yorkshire Terriers and Cavalier King Charles Spaniels, were reared from weaning to two years old on specific dietary regimens offering different levels of flavour and

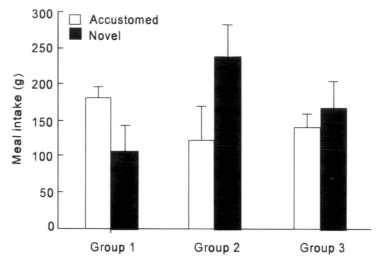

Figure 7.1 Intake of a food with novel flavour and texture characteristics relative to an accustomed food varies with previous food experience. Small and toy dogs fed a single nutritionally complete food for the first two years of life (Group 1) rejected the novel food relative to their accustomed food (t = -2.68, P < 0.05); whereas, dogs fed a variety of either commercially prepared dog foods (Group 2) or fresh and canned meat and fish (Group 3) accepted the novel food (t = 2.06, P > 0.5; t = 1.04, P > 0.05).

texture variety . When subsequently offered a novel food of unusual flavour and texture the dogs previously fed a variety of flavours preferred the novel food to their accustomed food, but dogs with restricted flavour experience (a commercially prepared puppy food) preferred their usual food (Figure 7.1). Hence the degree of previous experience of different flavours and textures is fundamental to the food selection behaviour of the individual.

Neophobia can also be induced through environmental conditions. The preference of adult cats for a familiar diet declined with repeated exposure to that diet, but testing in a novel environment resulted in a total rejection of the more novel food in favour of the familiar food (Figure 7.2). On return to their accustomed housing, the cats' preference reverted to the novel diet (Thorne, 1982).

AVERSION

Aversion develops rapidly to any food whose ingestion produces negative physiological responses. A nutritionally inadequate diet will often result in anorexia, for example, thiamine deficiency in cats (Everett, 1944), probably induced by a learned dietary aversion to the flavour of the inadequate food. The ability to associate a neophobia with a specific flavour is further evidence for a learned link between the food flavour and the physiological consequences of ingestion. Cats provided with a flavoured methionine deficient diet reduced their intake to about half requirement within 24 hours of first consumption of the food. Following development of this aversion, the cats rejected a nutritionally complete diet which included the same flavour (C.J. Thorne, unpublished data). Similarly administering lithium chloride during a single meal resulted in cats rejecting that food three days later and its effects could also be paired with the odour of the food alone (Mugford, 1977a). The development of flavour aversion is rapid and results in an immediate rejection of any toxic food, clearly an in-built safety feature. In the domestic dog, lithium chloride treatment is less effective, some dogs will eat the laced vomitus and the effect is often short lived. This lack of effect in the domestic dog may be due to reduced responsiveness caused through the process of domestication as long-lasting aversions have been demonstrated in wild coyotes (*Canis latrans*).

NOVELTY (NEOPHILIA)

The preference for new foods over a usual food is common in both cats and dogs, but is an apparent contradiction to neophobia. Novelty and neopho-

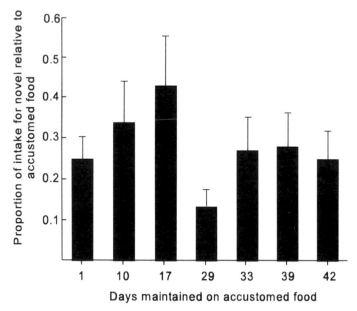

Figure 7.2 Neophobia to a relatively novel food can be induced through a change in the environment. Cats fed a single food show an increasing preference for a more novel food with increasing duration of single food feeding. Changing the test environment results in a rejection of the more novel food in favour of the accustomed food with return to the usual test environment reversing the preference.

bia complement each other with neophobia providing the means of avoiding potentially toxic foods and novelty providing the means for evaluating potential new food sources for nutritional quality and providing an alternative to the usual food if it becomes scarce. Which particular behaviour will be shown at a specific time will be dependent on context. A novel food whose characteristics are outside of the animal's feeding experience will tend to be rejected, whereas those within the known food set are accepted and tried. This variability in response is almost certainly due to the interaction between the duration of single food feeding and the palatability of the alternative food relative to the familiar food.

These three behavioural principles provide the animal with a sound base for learning about foods and developing a food selection strategy. They are adaptive in that they achieve avoidance of potentially dangerous foods, but wherever possible result in trial of potential new foods so that nutritional quality can be assessed. Aversion provides bottom line safety by ensuring that foods that produce negative physiological consequences do not become part of the food repertoire.

The role of variety

Overlying this behavioural basis of food selection are the individual differences in food likes and dislikes. There is some evidence that the various breeds respond differently to food, for example, Cavalier King Charles Spaniels and Labrador Retrievers appear to be particularly non-selective about food and the hounds show a food gorging ability equal to that of their wild ancestors. However, domestication, particularly for the dog, has produced a range of breeds with extreme differences in morphology and behaviour which would be expected to affect feeding behaviour and food selection. The same may also be true for the domestic cat although domestication and selective breeding have not resulted in the same extremes of differentiation as those between breeds of dogs. In addition, every domestic cat and dog is an individual, with unique genetic material and experience, which result in particular food likes and dislikes and a lesser or greater demand for variety within the diet. In the modern commercial pet food arena the customer has a wide range of formats, brands and varieties from which to choose and it is the role of the pet food manufacturer to provide nutritionally complete diets which meet the needs of the individual consumer.

In general, most domestic cats and dogs respond favourably to being provided with a variety of foods with the cat being more 'finicky' in its feeding habits than the dog. Flavour variety provides the owner with the means of maintaining their pet's interest in food. Feeding a variety of foods can increase intake in the short-term. Cats fed over three days on three meals of three different foods each day consumed a greater amount in terms of both weight and energy intake than when offered three meals of a single food (Mugford, 1977a). Clearly, just as in humans, sensory variety enhances the motivation to feed.

Summary

The ancestors of the domestic cat and dog developed as specialists to meet the demands of their ecological niches. This is evident in their anatomy, which is structured for capture of prey, and their dentition, which is characteristic of the carnivores with enlarged canine teeth for gripping and puncturing prey and the large carnassial teeth for cutting flesh. Similarly the senses of smell and taste are particularly sensitive to those chemicals that are relevant to the assessment of meat quality. This specialisation is inherited and passed from generation to generation, but in contrast the processes involved in food selection show a considerable degree of malleability with the repertoire of food items changing in response to food availability and quality. The learning processes involved in the development of food selection may well be genetically 'hard-wired' since both species appear to know what to learn and the cues which are relevant. For example, associations between a food flavour and the physiological consequence of food consumption are very rapidly learned, whereas it is very difficult to associate an auditory stimulus with the consequences of food intake.

Changes in the response to food items occur throughout life and provide an adaptive behavioural strategy which results in continual trial of new food items which will be added to the individual's set of acceptable foods. The greater the range of known foods the more likely the individual is to survive if the usual food items become scarce, thus providing competitive advantage.

For the young cat and dog, the first known safe solid food is provided by the parents. There is the possibility that food cues are provided through the mother's milk. When the young first start self-feeding they will tend to eat those foods which are also eaten by other adults. In this way they recognise foods that are available locally and this provides a safety factor for the young animal. As they mature and start to hunt their own food they will experiment and two contrasting behavioural strategies, neophobia and neophilia, provide the safety net to avoid very unusual foods and the tendency to seek variety respectively. For the domestic cat and dog these strategies still operate to a greater or lesser extent dependent on the individual and its previous experience, certainly most respond favourably to variety in their diet. Hence variety and palatability are important factors in the selection of nutritionally adequate foods, with food palatability the key driver in the longer term. Overall the domestic cat and dog show a very flexible strategy of food selection which has been developed in their wild ancestors.

References

Albone, E.S. (1984) Mammalian Semiochemistry: The Investigation of Chemical Signals Between Mammals, Chichester: John Wiley & Sons Limited

Baker, B.J., Booth, D.A., Duggan, J.P. and Gibson, E.L. (1987) Protein appetite demonstrated: Learned specificity of protein cue preference to protein need in adult rats. *Nutrition Research*, **7**, 481–487

Beauchamp, G.K., Maller, O. and Rogers, J.G. (1977) Flavor preferences in cats: effects on sucrose preference. *Science*, **171**, 699–701

Boudreau, J.C., Sivakumar, L., Do, L.T., White, T.D., Orovec, J. and Hoang, N.K. (1985) Neurophysiology of geniculate ganglion (facial nerve) taste systems: species comparisons. *Chemical Senses*, **10**, 89–127

Boudreau, J.C. (1989) Neurophysiology and stimulus chemistry of mammalian taste systems. In *Flavour Chemistry: Trends and Developments. American Chemical Society Symposium Series*, **388**, 122–137. Edited by R. Teranishi, R.G. Buttery and F. Shahidi

Bradshaw, J.W.S. (1986) Mere exposure reduces cats' neophobia to unfamiliar food. *Animal Behaviour*, **34**, 613–614

Bradshaw, J.W.S., Macdonald, D.W., Healey, L.M. and Arden-Clark, C. (1991) Differences in food preferences between populations and individuals of domestic cats *Felis catus. Behavioural Ecology.* Unpublished data

Burger, I.H. and Rivers, J.P.W. (1989) *Nutrition of the Cat and Dog* (Waltham Symposium No. 7). Cambridge: Cambridge University Press

Burghart, G.M. and Hess, E.H. (1966) Food imprinting in the snapping turtle (Cheldra serpentia). *Science*, **151**, 108–109

Campbell, R.G. (1976) A note on the use of a feed flavour to stimulate the feed intake of weaner pigs. *Animal Production*, **23**, 417–419

Carpenter, J.A. (1956) Species differences in taste preferences. *Journal of Comparative and Physiological Psychology*, **49**, 139–144

Denton, D.A. (1967) Salt appetite. In: *Handbook of Physiology, Vol. 1*, pp. 433–459. Edited by C.F. Code. Washington, D.C: American Physiological Society

Dodd, G.H. and Squirrel D.J. (1980) Structure and mechanism in the mammalian olfactory system. *Symposia of the Zoological Society of London*, **45**, 35–36

Domjan, M. (1973) Attenuation and enhancement of neophobia for edible substances. In: *Learning Mechanisms in Food Selection*, pp. 151–180. Edited by L.M. Barker, M.R. Best and M. Domjan. Waco, Texas: Baylor University Press

Everett, G.M. (1944) Observations on the behavior and neurophysiology of acute thiamine deficient cats. *American Journal of Physiology*, **141**, 439–448

Ewer, R.F. (1973) *The Carnivores*. London: Weidenfield & Nicolson

Ferrel, F. (1984a) Taste bud morphology in the fetal and neonatal dog. *Neuroscience and Biobehavioral Reviews*, **8(2)**, 175–183

Ferrel, F. (1984b) Gustatory nerve response to sugars in neonatal puppies. *Neuroscience and Biobehavioural Reviews*, **8(2)**, 185–190

Ferrel, F. (1984c) Effects of restricted dietary flavour experience before weaning on postweaning food preferences in puppies. *Neuroscience and Biobehavioural Reviews*, **8(2)**, 191–198

Fitzgerald, B.M. (1988) Diet of domestic cats and their impact on prey populations. In: *The Domestic Cat: the Biology of its Behaviour*, pp. 123–146. Edited by D.C. Turner and P. Bateson. Cambridge: Cambridge University Press

Fuchs, J.L. and Burghart, G.M. (1971) Effects of early feeding experience on the responses of garter snakes to food chemicals. *Learning and Motivation*, **2**, 271–279

Galef, B.G. and Henderson, P.W. (1972) Mother's milk: a determinant of the feeding preferences of weanling rat pups. *Journal of Comparative and Physiological Psychology*, **78**, 213–219

Hess, E.H. (1964) Imprinting in birds. *Science*, **146**, 1128

Houpt, K.A. (1978) Palatability and canine food preferences. *Canine Practice*, **5(6)**, 29–35

Houpt, K.A., Hintz, H.F. and Shepherd, P. (1978) The role of olfaction in canine food preferences. *Chemical Senses and Flavour*, **3**, 281–290

Kalmus, H. (1955) The discrimination by the nose of the dog of individual human odours and in particular of the odours of twins. *British Journal of Animal Behaviour*, **3**, 281–290

Kitchel, R.L. (1978) Taste perception and discrimination by the dog. *Advances in Veterinary Science and Comparative Medicine*, **22**, 287–314

Kuo, Z.Y. (1967) *The Dynamics of Behaviour Development: An Epigenetic View.* New York: Random House

LeMagnen, J. (1967) Habits and food intake. In: *Handbook of Physiology, Vol. 1*, pp. 11–30. Edited by C.F. Code. Washington, D.C.: American Physiological Society

Ling, E.R., Kan, S.K. and Porter, J.W.G. (1961) The composition of milk and the nutritive value of its components. In *Milk: The Mammary Gland and its Secretion. Vol II.* Edited by S.K. Kan & A.T. Cowrie. New York: Academic Press

Lohse, C.L. (1974) Preferences of dogs for various meats. *Journal of the American Animal Hospital Association*, **10**, 187–192

MacDonald, M.L., Rogers, Q.R. and Morris, J.G. (1984) Nutrition of the domestic cat, a mammalian carnivore. *Annual Review of Nutrition*, **4**, 521–562

MacDonald, M.L., Rogers, Q.R. and Morris, J.G. (1985) Aversion of the cat to dietary medium-chain triglycerides and caprylic acid. *Physiology and Behavior*, **35**, 371–375

Minstretta, C.M. (1972) Topographical and histological study of developing rat tongue, palate and taste buds. In *The Third Symposium on Oral Sensation and Perception: The Mouth of the Infant*, pp. 11–30. Edited by J.F. Bosma. Springfield: Thomas

Minstretta, C.M. and Bradley, R.M. (1977) Taste in utero: theoretical considerations. In *Taste and Development – The Genesis of Sweet Preference.* Edited by. J.M. Weiffenbach. Bethesda, Maryland: DHEW Publications

Morrison, G. R. (1974) Alterations in palatability of nutrients for the rat as a result of prior testing. *Journal of Comparative and Physiological Psychology*, **86**, 56

Mugford, R.A. (1977a) External influences on the feeding of carnivores. In *The Chemical Senses and Nutrition*, pp. 25–50. Edited by M.R. Kare & O. Maller. New York: Academic Press

Mugford, R.A. (1977b) Comparative and developmental studies of feeding in dogs and cats. *British Veterinary Journal*, **133**, 98

National Research Council (1985) *Nutrient Requirements of Dogs.* Washington, D.C.: National Academy of Sciences

National Research Council (1986) *Nutrient Requirements of Cats.* Washington, D.C.: National Academy of Sciences

Rozin, P. (1967) Thiamine specific hunger. In: *Handbook of Physiology, Vol. 1*, pp. 411–431. Edited by C.F. Code. Washington, D.C.: American Physiological Society

Rozin, P. and Kalat, J.W. (1971) Specific hungers and poison avoidance as adaptive specialisations of learning. *Psychological Reviews*, **78**, 459–486

Scott, E. M. (1946) Self-selection of diet. Appetite for protein. *Journal of Nutrition*, **32**, 293–301

Scott, E. M. and Quint, E. (1946) Self selection of diet. III. Appetites for B vitamins. *Journal of Nutrition*, **32**, 285–291

Smotherman, W.P. (1982) In utero chemosensory experience alters taste preferences and corticosterone responsiveness. *Behavioural and Neural Biology*, **36**, 61–68

Soulairac, A. (1967) Control of carbohydrate intake. In: *Handbook of Physiology*, *Vol. 1*, pp. 387–398. Edited by C.F. Code. Washington, D.C.: American Physiological Society

Thorne, C.J. (1982) Feeding behaviour in the cat – recent advances. *Journal of Small Animal Practice*, **23**, 555–562

White, T.D. and Boudreau, J.C. (1975) Taste preferences in the cat for neurophysiologically active compounds. *Physiological Psycholology*, **3**, 405–410

8

RECENT FINDINGS ON THE EFFECTS OF NUTRITION ON THE GROWTH OF SPECIFIC BROILER CARCASS COMPONENTS

A.W. WALKER, J. WISEMAN[1], N.J. LYNN AND D.R. CHARLES
ADAS Gleadthorpe, Meden Vale, Mansfield, Nottinghamshire, NG20 9PF, UK
[1] *University of Nottingham, Sutton Bonington Campus, Loughborough, Leicestershire, LE12 5RD, UK*

Introduction

During the early years of the development of the broiler industry, nutrition research was aimed almost entirely at improving growth rate and food conversion efficiency. In today's poultry industry this alone is no longer sufficient.

UK poultry meat consumption has increased from 14.5 kg/person/year in 1982 to 20.6 kg/person/year in 1993 (MAFF, 1994), partly at the expense of red meat. Added value products, including breadcrumbed and flavour-coated products available from the expanding take-away or 'fast-food' sector, account for an increasing proportion of total consumption (British Chicken Information Service, 1994). This is not surprising since poultry meat lends itself well to processing, marinating and manufacturing. The growing catering trade in the UK now accounts for 27% of family expenditure on food, and this proportion is increasing (MAFF, 1991). This trend has reflected the 46% increase in the number of restaurants and cafes in the UK in the period from 1982 to 1991 (Central Statistical Office, 1992).

These newly expanding markets demand mainly white breast meat, and modern nutrition research and practice must aim therefore to maximise usable meat yield according to the particular market for which the bird is intended.

Many attempts have been made to express the growth of animals in mathematical terms. Although varying in specific details, these models have a sigmoid form. Of several functions reviewed by Wilson (1977), the Gompertz function best described broiler chicken growth. This function (Figure 8.1) describes the likely point of maximum growth rate (the point of inflection of the curve) and the likely mature bodyweight (the asymptote).

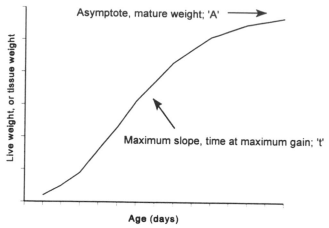

Figure 8.1 Growth as characterised by the Gompertz equation.

For poultry the best known growth models are probably those of Emmans (1981), which modelled broiler lean tissue growth, allowing estimation of the yield of carcass parts, and the Reading model (see Fisher, 1983), which offered a theoretical approach to the assessment of amino acid intake effects, using lysine as the limiting factor.

A practical method of optimising protein intake in order to improve growth rate and lean tissue deposition was described by Filmer (1993) who pointed out that it is necessary to provide an appropriate intake for the age of the bird, independently of the effects on voluntary intake of either dietary energy content or environmental temperature. This model and others are likely to be good predictors of protein deposition but their relevance to modern market needs could be improved by incorporating more information on the anatomical distribution of growth among carcass parts.

This chapter provides data on some effects of nutrition on the growth rate of broilers, the composition of broiler carcasses, and on the rate of deposition of tissue for specific carcass components at different stages of the bird's growth. The data have been accumulated from a number of experiments which were done at both ADAS Gleadthorpe and the University of Nottingham in the early to mid-1980s. The authors acknowledge that genetic selection for increased lean meat yield has continued since that time, but they propose that there is no reason why the nutritional effects on the carcass should not hold true for the modern broiler. Accordingly it was considered appropriate to submit the work to new analysis in order to better exploit it for current needs.

Table 8.1   DETERMINED ANALYSES OF BASAL GLEADTHORPE DIETS (%)

	Diet			
	H1	*H2*	*H3*	*H4*
Dry matter	88.10	87.60	88.30	88.30
Crude protein	25.90	24.10	22.60	21.20
Lysine	1.28	1.23	1.16	1.02
Methionine	0.54	0.49	0.51	0.45
Cystine	0.40	0.38	0.38	0.36
ME(MJ/kg)	14.10	13.90	14.30	14.10
	L1	*L2*	*L3*	*L4*
Dry matter	89.40	87.60	89.20	88.50
Crude protein	22.50	21.60	20.90	19.10
Lysine	1.04	1.00	0.95	0.83
Methionine	0.50	0.43	0.40	0.39
Cystine	0.40	0.38	0.35	0.36
ME (MJ/kg)	12.70	12.90	12.90	13.10
	HM1	*HM2*	*HM3*	*HM4*
Dry matter	88.40	87.40	87.90	88.50
Crude protein	24.60	23.10	22.10	20.40
Lysine	1.25	1.14	1.02	0.98
Methionine	0.51	0.48	0.46	0.42
Cystine	0.39	0.41	0.39	0.37
ME(MJ/kg)	13.90	13.70	13.90	13.70
	LM1	*LM2*	*LM3*	*LM4*
Dry matter	88.10	86.80	88.90	88.50
Crude protein	24.10	22.10	21.20	19.60
Lysine	1.24	1.15	1.03	0.88
Methionine	0.51	0.46	0.43	0.41
Cystine	0.42	0.40	0.36	0.34
ME(MJ/kg)	13.50	13.40	13.50	13.50

Data generated from Gleadthorpe

The work at Gleadthorpe Research Centre (ADAS, unpublished data) was designed to investigate the effects of a wide range of feeding programmes for male roaster chickens (i.e. broilers grown to heavy weights), on body weight gain, feed conversion efficiency and carcass composition at a particular age.

A total of 16 diets (Table 8.1) were formulated, and these were designed to give a range of 4 energy densities, described as High (H), High-Medium (HM), Low-Medium (LM) and Low (L). In each category there were four energy density:lysine (MJ:% lysine) ratios, fixed at a level similar to that found in commercial broiler diets. These ratios were described as 1,2, 3 and 4 (see Table 8.2). In some treatments the same ratio was fed throughout (for

Table 8.2　SUMMARY OF DIETS AND GLEADTHORPE DIET CODES

Energy density	Energy lysine ratio	Diet code
High (13.1 MJ/kg)	10.0	H1
	11.2	H2
	12.3	H3
	13.8	H4
High-medium (12.5 MJ/kg)	10.1	HM1
	11.2	HM2
	12.3	HM3
	13.8	HM4
Low-medium (12.0 MJ/kg)	10.1	LM1
	11.2	LM2
	12.4	LM3
	13.7	LM4
Low (11.4 MJ/kg)	10.2	L1
	11.2	L2
	12.4	L3
	13.7	L4

Table 8.3　SUMMARY OF TREATMENTS (GLEADTHORPE)

Treatment	Day 0–9	Day 9–21	Day 21–42	Day 42–70
1111 Group				
1	H1	H1	H1	H1
2	HM1	HM1	HM1	HM1
3	LM1	LM1	LM1	LM1
4	L1	L1	L1	L1
1234 Group				
5	H1	H2	H3	H4
6	HM1	HM2	HM3	HM4
7	LM1	LM2	LM3	LM4
8	L1	L2	L3	L4
2234 Group				
9	H2	H2	H3	H4
10	HM2	HM2	HM3	HM4
11	LM2	LM2	LM3	LM4
12	L2	L2	L3	L4
3334 Group				
13	H3	H3	H3	H4
14	HM3	HM3	HM3	HM4
15	LM3	LM3	LM3	LM4
16	L3	L3	L3	L4
4444 Group				
17	H4	H4	H4	H4
18	HM4	HM4	HM4	HM4
19	LM4	LM4	LM4	LM4
20	L4	L4	L4	L4

Table 8.4   THE EFFECT OF ENERGY:LYSINE RATIO ON THE BODYWEIGHT OF
BROILERS (KG) (GLEADTHORPE)

| | Treatment | | | | | | | |
Age(d)	*1111*	*1234*	*2234*	*3334*	*4444*	*mean*	*sed ±*	*p*
DO	0.04	0.04	0.04	0.04	0.04	0.04	0.000	NS
9	0.21	0.21	0.20	0.18	0.15	0.19	0.004	***
21	0.76	0.74	0.71	0.63	0.50	0.67	0.011	***
28	1.25	1.22	1.16	1.07	0.87	1.11	0.020	***
35	1.77	1.72	1.66	1.56	1.33	1.61	0.027	***
42	2.31	2.23	2.20	2.10	1.87	2.14	0.029	***
49	2.84	2.79	2.76	2.63	2.47	2.70	0.031	***
56	3.24	3.21	3.19	3.09	2.97	3.14	0.036	***
63	3.64	3.70	3.62	3.54	3.39	3.58	0.069	***
70	3.98	3.93	3.97	3.93	3.71	3.90	0.052	***

example, in programme 1111) and in some treatments the ratio was
changed periodically (for example in programme 2234). The diets were fed
in stages; i.e., starter, grower, finisher 1 and finisher 2 (roaster) phases.

The combination of diets available meant that 20 treatments could be
used, and depending on the precise treatment, changes were made at one or
more of the following ages: 9, 21, 42 days (Table 8.3). Food was available to
the birds *ad libitum*.

Bird bodyweight, feed usage and feed conversion efficiency were
measured at 9, 21, 42 and 70 days of age. Carcass yield data for the
Gleadthorpe work are available only for birds at 70 days of age, at which
time a sample of six birds per treatment was taken and, in addition to live
weight, weights of eviscerated carcass, breast meat, thigh drumstick weight
and abdominal fat pad were recorded.

Significant bodyweight differences were achieved throughout the growing period, and these were evident as early as 9 days of age (Table 8.4). The
pattern which was established at this early stage was for birds receiving diets

Table 8.5   THE EFFECT OF ENERGY: LYSINE RATIO ON THE FEED USAGE OF
BROILERS (G/BIRD/DAY) (GLEADTHORPE)

| | Treatment | | | | | | | |
Period (d)	*1111*	*1234*	*2234*	*3334*	*4444*	*mean*	*sed ±*	*p*
0.9	24.3	23.5	23.7	22.0	20.9	22.9	0.491	***
0–21	52.6	51.9	50.5	45.9	37.2	47.6	0.652	***
0–42	96.3	96.2	94.8	91.0	81.9	92.0	1.319	***
0–63	118.7	118.1	117.5	114.9	108.3	115.5	1.549	***
9–21	74.4	73.8	71.0	64.0	49.7	66.6	1.045	***
21–42	141.9	141.7	140.5	137.2	128.1	137.9	2.290	***
42–63	180.9	178.5	180.3	180.9	179.9	180.1	3.087	NS
63–70	192.2	186.2	194.5	198.2	192.8	192.8	5.520	NS

Table 8.6   THE EFFECT OF ENERGY-LYSINE RATIO ON FEED CONVERSION
EFFICIENCY OF BROILERS (GLEADTHORPE)

Period(d)	1111	1234	2234	3334	4444	mean	sed ±	p
			Treatment					
0–92	0.775	0.791	0.708	0.663	0.558	0.699	0.015	***
0–21	0.642	0.631	0.621	0.606	0.571	0.532	0.007	***
0–42	0.545	0.528	0.532	0.530	0.524	0.532	0.004	***
0–70	0.426	0.421	0.429	0.431	0.428	0.427	0.005	NS
9–21	0.609	0.592	0.599	0.591	0.579	0.594	0.008	*
21–42	0.508	0.489	0.498	0.503	0.510	0.502	0.005	**
42–70	0.292	0.302	0.318	0.326	0.336	0.315	0.011	**

in the 1111 and 1234 groups – i.e. a narrow energy:lysine ratio fed
throughout and up to 42 days respectively – to be significantly heavier than
those receiving diets in the 2234, 3334 and 4444 groups. These latter
groups, in which the energy:lysine ratio became progressively wider as the
birds aged (or started at a wide ratio and continued as such in the case of the
4444 group) gave lighter birds at each weighing occasion, although the
magnitude of any differences reduced as a proportion of total bodyweight as
the birds grew older. At 70 days of age only those grown on the 4444
programme were significantly lighter than the other treatments. Variations
in feed intake are reported in Table 8.5.

Differences in feed conversion efficiency (Table 8.6) were apparent at 9
days of age and reflected the bodyweight differences noted earlier. Feeding
programmes which started with a narrow energy:lysine ratio (1111 and
1234) produced the best efficiencies at this early age ($p < 0.001$) and this
difference was sustained at 42 days of age. However, over the period 0–70
days there were no significant effects of the feeding programmes on conversion efficiency, reflecting the fact that the wide energy:lysine groups
(2234, 3334 and 4444) achieved superior performance in the 42–70 day
period and effectively nullified the conversion efficiency advantage of the
1111 and 1234 groups in the period up to 42 days. Importantly, the least
efficient programme in the latter growing stage was programme 1111, and

Table 8.7   THE EFFECT OF ENERGY : LYSINE RATIO ON MORTALITY OF
BROILERS (%)

Period(d)	1111	1234	2234	3334	4444	mean	sed ±	p
			Treatment					
0–21	2.1	2.7	2.1	1.5	2.4	2.2	1.19	NS
0–42	9.6	8.4	6.6	5.7	6.0	7.3	1.98	NS
0–70	15.9	16.0	11.5	9.0	7.2	11.9	2.71	**

Table 8.8   THE EFFECT OF ENERGY: LYSINE RATIO ON THE CARCASS
COMPOSITION OF BROILERS, 70 DAYS (GLEADTHORPE)

Component	Treatment					mean	sed ±/-	p
	1111	*1234*	*2234*	*3334*	*4444*			
Eviscerated carcass (g)	2907.0	2739.0	2880.0	2627.0	2621.0	2755.0	74.5	***
Abdominal fat/ eviscerated carcass (%)	3.2	4.5	3.7	4.2	4.1	3.9	0.41	*
Breast meat/ eviscerated carcass (%)	22.9	21.9	22.7	21.7	21.0	22.1	0.48	**
Leg meat/eviscerated carcass (%)	30.0	30.1	29.7	29.9	30.6	30.1	0.51	NS

the best was 4444 – a complete reversal of the position in the early phase of
the experiment.

A significant treatment effect ($p < 0.01$) on mortality was observed over
the full growing period (Table 8.7) and this highlighted the possible detrimental effect of rapid early growth on young broilers. There was a progress
ive reduction in mortality when birds were fed diets with a wider energy:
lysine ratio, and this trend had established itself by 42 days, albeit not
significantly. The reduction in mortality was in line with the trends for
lighter bodyweights which occurred on programmes 2234, 3334 and 4444.

The carcass composition data (Table 8.8) show some similarities with the
trends found in the growth data described above. In terms of breast meat
yield, a significant, though small, increase was obtained when regime 1111
was fed compared with the other treatments. The lowest breast meat yield
was obtained on regimes 3334 and 4444, again indicating that protein
requirements were not being met when feeding wide energy:lysine diets
early in the life of the bird. Although significantly lower in breast meat yield,
these treatments gave the same leg meat yield as the narrow ration regime,
but had significantly more abdominal fat as a proportion of eviscerated
carcass yield. The increased abdominal fat content of the carcass is likely to
be the result of excessive energy intake as the birds consumed more feed in
an attempt to meet amino acid requirements.

Modelling of responses

The influence of nutrition on overall body and carcass weight, together with
estimates of the degree of fatness, has been the topic of many studies.
However, most programmes have been concerned only with the situation

Table 8.9   B, M, C AND A ESTIMATES FROM THE GOMPERTZ EQUATION FOR
LIVE WEIGHT AND FEED INTAKE (GLEADTHORPE)

	Diet				
	1111 *(A)*	*1234* *(B)*	*2234* *(C)*	*3334* *(D)*	*4444* *(E)*
Body Weight					
B	0.0410	0.0402	0.0395	0.0380	0.0443
M	36.0	37.2	38.4	40.7	41.3
C	5132	5216	5323	5476	4872
A	-36.1	-24.6	-24.2	-20.8	53.6
P[a]	100.0	99.9	100.0	100.0	100.0
Feed Intake					
B	0.0237	0.0253	0.0239	0.0243	0.0269
M	58.5	55.6	58.8	60.0	59.3
C	20073	18481	19940	19782	17650
A	-606	-530	-562	-465	-214
P	99.9	99.9	99.8	99.8	99.8

[a] Proportion of variance accounted for by the model fitted

at slaughter, or at diet changeover between hatching and slaughter. Such
an approach does not allow accurate estimates of growth and development
of live weight and carcass components over time. However, such analyses
are of critical importance to decisions on nutritional regimes and optimum
time of slaughter. Furthermore these influences on carcass components
rather than entire live weight are assuming more importance as portions
assume increasing importance in the poultry meat market. Accordingly, it is
possible that managerial decisions on nutritional regime and optimum time
of slaughter could vary depending upon which carcass component is

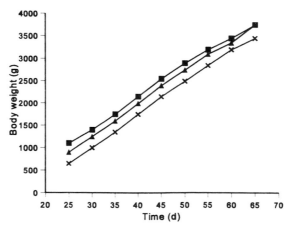

Figure 8.2 Growth of body weight (Gleadthorpe) ■ A, ▲ C, X E.

considered and the relative financial value of the individual carcass components.

Modelling of growth in poultry has received considerable attention although a critical appraisal of the models utilised is considered outside the scope of this chapter. However, Wilson (1977) was of the opinion that the Gompertz equation (e.g., Laird, Tyler and Barton, 1965) was a suitable function to employ. The characteristic sigmoid shape of this function is presented in Figure 8.1 which also indicates the two principal parameters fitted, being mature weight (which, whilst mathematically important, is perhaps of minor significance as broilers are always slaughtered well before this) and time at maximum rate of growth. The model has been utilised with poultry on many occasions (e.g. Tzeng and Becker, 1981; Pasternak and Shalev, 1983; Ricklefs, 1985; Anthony, Emmerson, Nestor and Bacon, 1991; Knizetova, Hyanek, Knize and Roubicek, 1992) although studies have been confined usually to whole carcass weight.

In the current studies where body weights and carcass components were analyzed using the Gompertz curve of the form:

$$W = A + C * \exp * (-\exp(-B(t-M)))$$
W   =   weight at time, t, of live body weight or carcass component
t   =   age of chicken (d)

The parameters A + C, B and M were interpreted as follows:

A + C =   the asymptotic weight approached, an estimate of the mature weight

B   =   the rate of exponential decay of the initial growth rate, a measure of the decline in growth rate

M   =   the age at which growth is maximum

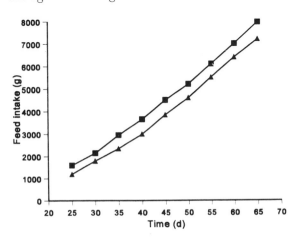

Figure 8.3 Increase in feed intake (Gleadthorpe) ■ A, ▲ E.

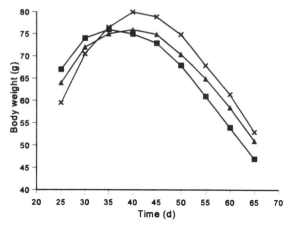

Figure 8.4 Rate of growth of body weight (Gleadthorpe) ■ A, ▲ C, X E.

The data generated from Gleadthorpe for body weight and food intake on a weekly basis allowed the fitting of Gompertz models. Data for A + C, B and M, together with the proportion of the variance in data accounted for by the models fitted, are given in Table 8.9 and responses obtained are presented in Figures 8.2 and 8.3, for treatments A, C and E, and Figure 3, for treatments A and E, respectively over the time period 25 to 65 days of age in 5 day increments. Further analysis of the data involved differentiating the Gompertz functions to obtain dW/dt (i.e. rate of gain of live weight or carcass component) over time and the results for such an analysis are given in Figures 8.4 and 8.5 respectively for live weight and feed intake. The data indicate that the more nutritionally deficient regime (E) was

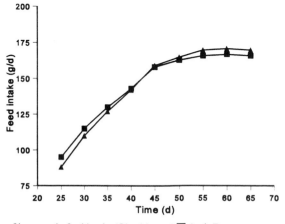

Figure 8.5 Rate of increase in feed intake (Gleadthorpe) ■ A, ▲ E.

Table 8.10   NUTRITIONAL VALUE OF NOTTINGHAM DIETS

	Starter		Finisher	
	Nutrient Concentration[a]			
	High (H)	*Low (L)*	*High (H)*	*Low (L)*
Calculated Analysis:				
Protein (g/kg)	247.80	199.40	220.30	182.90
AME (MJ/kg)	14.15	11.39	14.29	11.85
Protein:energy				
(g/kg:MJ ME/kg)	17.51	17.51	15.44	15.43
Methionine (g/kg)	5.00	4.00	4.50	3.80
Lysine (g/kg)	14.50	11.70	2.50	10.40

[a] Commercial' (C) starter and finisher diets were prepared by mixing equal weights of the respective H and L diets

Table 8.11   B, M, C AND A ESTIMATES FROM THE GOMPERTZ EQUATION FOR LIVE WEIGHT AND SELECTED CARCASS COMPONENTS (NOTTINGHAM)

	Diet				
	H-H	*H-L*	*C*	*L-H*	*L-L*
Body Weight (g)					
B	0.0553	0.0337	0.0353	0.0523	0.0319
M	29.2	40.1	39.8	38.0	423.7
C	3585	5067	5241	4234	4603
A	11.0	-84.0	-71.0	-80.8	13.3
P[a]	95.5	99.6	99.8	99.9	99.8
Breast Muscle (g)					
B	0.0395	0.0209	0.0290	0.0516	0.0395
M	38.3	65.1	51.6	42.4	46.8
C	846	1606	1211	747	772
A	-3.2	-28.2	-8.5	17.2	10.4
P	99.5	99.6	99.9	99.8	99.3
Thigh/Leg Muscle (g)					
B	0.0395	0.0209	0.0290	0.0516	0.0395
M	33.9	53.0	45.5	41.6	44.6
C	824.9	1200.0	1009.5	783.1	757.0
A	-6.4	-24.7	-8.9	11.5	7.8
P	99.5	99.6	99.9	99.8	99.3
Abdominal Fat Pad (g)					
B	0.0468	0.0416	0.0331	0.0423	0.0554
M	43.3	42.7	53.4	50.4	43.8
C	114.3	64.6	135.6	127.1	52.3
A	0.17	-0.05	-1.15	-0.03	0.96
P	99.2	98.0	98.7	99.7	93.9

[a] Percentage variance in data accounted for by fitted model

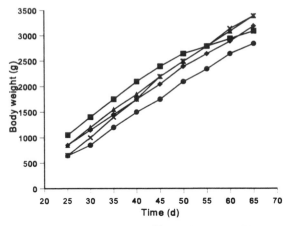

Figure 8.6 Growth of body weight (Nottingham) ■ HH, ◆ HL, ▲ C, X LH, ● LL.

associated with lower feed intakes and, although there was a higher maximum rate of growth of bodyweight, the rate of increase and decrease in this rate of growth was such that birds on E were never able to achieve the live bodyweights of those on A.

Studies of the growth of live weight of broilers as influenced by nutritional regimes were investigated in a trial (University of Nottingham, unpublished data) which also examined specific carcass components. Birds were fed a series of starter and finisher diets (with diet changeover at 21 days of age). Starter diets had crude protein (CP) contents of 248 (High – H) and 199g/kg (Low – L) and corresponding apparent metabolisable energy (AME) values were 14.15 and 11.39 MJ/kg respectively. Protein:energy ratios were therefore maintained at 17.5. The CP contents of finisher diets

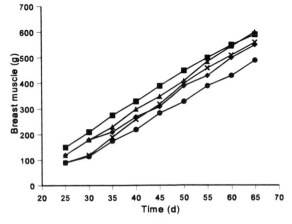

Figure 8.7 Growth of breast muscle (Nottingham) ■ HH, ◆ HL, ▲ C, X LH, ● LL.

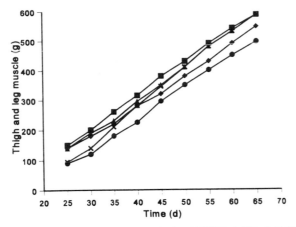

Figure 8.8 Growth of thigh and leg muscle (Nottingham) ■ HH, ◆ HL, ▲ C, X LH, ● LL.

were 220 (H) and 183 (L) g/kg, AME values were 14.29 (H) and 11.85 (L) MJ/kg with corresponding protein:energy ratios at 15.4. A third diet in both the finisher and starter series (C) was manufactured by mixing equal amounts of H and L. Nutritional values for C were therefore intermediate between H and L (Table 8.10). Representative samples of birds were slaughtered each week up to 10 weeks of age and dissected, using a standard procedure, into component tissues.

The fitted parameters for live weight, breast meat ('white' meat), thigh + leg meat ('dark meat') and abdominal fat pad are presented in Table 8.11 and fitted curves based on increments of 5 days, from day 25, in Figures 8.6, 8.7, 8.8, and 8.9 respectively. Further responses are presented

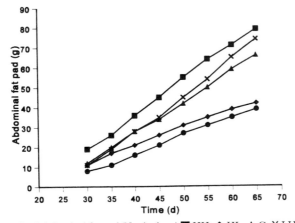

Figure 8.9 Growth of abdominal fat pad (Nottingham) ■ HH, ◆ HL, ▲ C, X LH, ● LL.

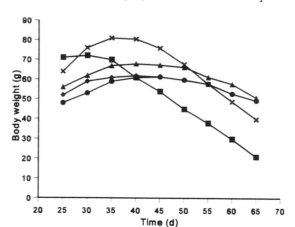

Figure 8.10 Rate of growth of body weight (Nottingham) ■ HH, ♦ HL, ▲ C, X LH, ● LL.

in Figures 8.10, 8.11, 8.12 and 8.13 respectively for rate of growth of live
weight, breast muscle, thigh/leg muscle and abdominal fat pad.

Response curves confirm that diets of high nutrient concentration promote more rapid live weight gain but that this is accompanied by increasing
amounts of carcass fat (abdominal fat pad is a minor component of carcass
fat but is a reasonably good predictor of it). It is evident, however, that
altering nutritional regime has a pronounced effect on the rate of gain of
both body weight and individual carcass components with the confirmation
of the observation that early feed restriction (i.e feeding the 'L' diet during
the starter phase) is followed by a 'compensatory' period when diets of
higher nutrient concentration (i.e. diet 'H' in the finisher phase) are offered

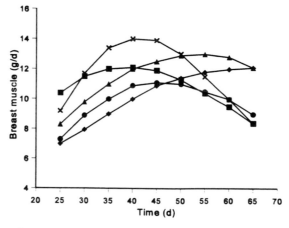

Figure 8.11 Rate of growth of breast muscle (Nottingham) ■ HH, ♦ HL, ▲ C, X LH, ● LL.

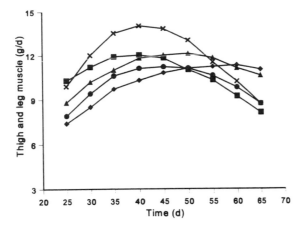

Figure 8.12 Rate of growth of thigh and leg muscle (Nottingham) ■ HH, ◆ HL, ▲ C, X LH, ● LL.

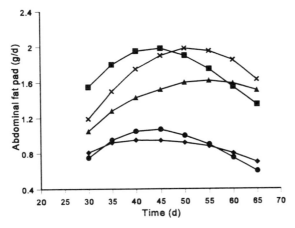

Figure 8.13 Rate of growth of abdominal fat pad (Nottingham) ■ HH, ◆ HL, ▲ C, X LH, ● LL.

(e.g. Hargis and Creger, 1980; Plavnik and Hurwitz, 1985; Jones and Farrell 1992).

It is apparent that the 'H-H' regime invariably results in greater overall weight of individual carcass meat components until around 50 days of age but that the rate of gain is lower than with the 'L-H' combination. The growth of fat tissue is always greater in 'H-H' over the entire growth period evaluated. In contrast to the 'whole-bird' market where live weight is of paramount importance (irrespective of carcass composition, i.e. carcass fat will contribute to overall weight) the 'portioning' market is dependent upon the weight of the individual component in question and the fat content is of much less relevance. Accordingly the contribution of growth of fat within the carcass in this latter case will be less significant.

Further approaches to this subject could involve the modelling of responses through manipulation of amino acid balance and protein:energy ratio together with comparisons between genotypes and generations. Invariably a consideration of the relative price of the dietary regimes employed will be of fundamental importance.

Acknowledgement

Funding for part of this work was from the Ministry of Agriculture Fisheries and Food whose support is gratefully acknowledged.

References

Anthony, N.B., Emmerson, D.A., Nestor, K.E. and Bacon, W.L. (1991) *Poultry Science*, **70**, 13–19

British Chicken Information Service (1994) Review of the British Retail Chicken Market

Central Statistical Office (1992) United Kingdom Statistics. HMSO

Emmans, G.C. (1981) A model for the growth and feed intake of ad libitum fed animals, particularly poultry. In *British Society of Animal Production Occasional Publication*, **5**, 103–110

Filmer, D.G. (1993) *Proceedings of the 8th International Poultry Breeders' Conference*, Glasgow, 25–26

Fisher, C (1983) *Turkeys*, **31**, 39–50

Hargis, P.H. and Creger, C.R. (1980) *Poultry Science*, **59**, 1499–1504

Jones, G.P.D. and Farrell, D.J. (1992) *British Poultry Science*, **33**, 579–587

Knizetova, H., Hyanek, B., Knize, B. and Roubicek, J. (1992) *British Poultry Science*, **32**, 1027–1038

Laird, A.K., Tyler, S.A. and Barton, A.D. (1965) *Growth*, **29**, 233–248

Ministry of Agriculture Fisheries and Food (1991) Household Food Consumption and Expenditure. HMSO

Ministry of Agriculture Fisheries and Food (1994) Agriculture in the United Kingdom. HMSO

Plavnik, I. and Hurwitz S. (1985) *Poultry Science*, **64**, 348–355

Pasternak, H. and Shalev, B.A. (1983) *British Poultry Science*, **24**, 531–536

Ricklefs, R.E. (1985) *Poultry Science*, **64**, 1563–1576

Tzeng, R. and Becker, W.A. (1981) *Poultry Science*, **60**, 1101–1106

Wilson, B.J. (1977) In *Growth and Poultry Meat Production* pp. 89–115. Edited by K.N. Boorman and B.J. Wilson. British Poultry Science Ltd, Edinburgh.

IV

Pig Nutrition

ERRATUM

Table 9.3 PREDICTION OF DC_e, MC_e OR DE BY STEP-WISE REGRESSION ANALYSIS BASED ON THE CHEMICAL COMPOSITION OF INDIVIDUAL FEEDINGSTUFFS.

Intercept slope/regressor	R^2	CV (%)
(1) $DC_e = 0.867-0.029_{xylose}$	0.99	3.2
(2) $MC_e = 0.822-0.034_{xylose}$	0.97	7.8
(3) $DE=15.31-0.093_{NDF}+0.269_{oil}-0.357_{ash}-0.124_{xylose}+0.08_{UAC}$	0.86	10.0
(4) $DE=16.69-0.145_{NDF}-0.289_{ash}$	0.94	8.7
(5) $DE=12.32+0.259_{oil}-0.618_{ash}+0.11_{nitrogen}+0.158_{glucose}-0.1_{NSP}$	0.95	5.6
(6) $DE=11.576+0.052_{starch}-0.066_{oil}-0.418_{UAC}+0.327_{NSP}-0.48_{glucose}$	0.99	1.8

UAC = uronic acids

Equations 1 and 2 are based on 8 alternative feedstuffs which had been both chemically analysed and for which DC_e and MC_e had been determined *in vivo*. Equation 3 is based on 32 alternative feedstuffs which had been chemically analysed, but *in vivo* measurements of DE had not been obtained for 23 of them; for these the mean of the range of published values was used. Equation 4 is based on values for 13 graminaceous by-products, equation 5 on 11 leguminous samples and equation 6 is based on values for by-products from 8 non-legume dicotyledonous species.

Table 9.5 PREDICTION OF DC_e AND MC_e BY STEP-WISE REGRESSION BASED ON CHEMICAL CONSTITUENTS OF 28 DIETS

Intercept Slope/Regressor	R^2	CV(%)
(7) $DC_e = 0.966 - 0.03485_{xylose}$	0.80	1.4
(8) $DC_e = 0.9806 - 0.0283_{xylose} - 0.00516_{glucose}$	0.92	1.2
(9) $MC_e = 0.944 - 0.03404_{xylose}$	0.79	2.6
(10) $MC_e = 0.955 - 0.02863_{xylose} - 0.00424_{glucose}$	0.90	1.9

FOOTNOTE TO **TABLE 9.7**

F = samples incubated in a faecal inoculum for 48 h according to Löwgren *et al.,* (1989),
D = samples incubated in a duodenal inoculum for 12 h according to Löwgren *et al.,* (1989),
% ferm is a measure of the fermentation of samples (%ferm = F=D/FX100)

Equation 11 is based on nine alternative feeds containing substantial amounts of NSP, but little starch. Equations 12 and 13 are based on the addition of a further three NSP-rich alternative feeds which also contained from 150 to 200g starch/kg DM.

9

PREDICTION OF THE ENERGY VALUE OF ALTERNATIVE FEEDS FOR PIGS

A.C. LONGLAND AND A.G. LOW
Institute of Grassland and Environmental Research, Plas Gogerddan, Aberystwyth, Dyfed, SY23 3EB, UK

Introduction

In recent years much attention has been focused on developing systems to determine the energy value of pig feeds, because the efficiency with which pigs digest and utilise dietary energy is a primary determinant of the extent to which a feedstuff can be converted into useful products. In conventional pig feeding systems, cereal starch forms the major source of dietary energy. However, there is increasing interest in using alternatives and many of plant origin contain substantial amounts of non-starch polysaccharides (NSP) and low or intermediate levels of starch and soluble sugars.

This chapter describes briefly the chemical composition and digestion of NSP and starch by pigs, and summarises the methods by which the energy value of feeds is assessed *in vivo*. Various laboratory-based protocols which may be used to predict the energy value of alternative feedstuffs for pigs are also discussed.

Chemical composition of NSP

NSP, together with lignin are equivalent to dietary fibre (Trowell, Southgate, Wolever, Leeds, Gasull and Jenkins, 1976). The NSP of the plant cell wall consists of structural or storage polysaccharides. The major monomeric NSP constituents are the 6-deoxyhexose sugar rhamnose, the pentoses arabinose and xylose, the hexose sugars glucose, galactose and mannose and the hexuronic acids, glucuronic or galacturonic acids. Glucose in NSP is largely derived from the cellulosic portion, whereas xylose, mannose and some galactose come from hemicellulose. The uronic acids and much of the arabinose and galactose originate from the pectic fraction. The quantity of each constituent, and their physical and chemical associations with each other and other cell wall components differs widely between species, between plant parts within a species and between cell types within a plant part. During cell division the middle lamella, which is rich in pectins, is the first cell wall structure to be laid down and serves to separate the dividing cells. The primary cell wall which is deposited inside the middle lamella contains cellulose microfibrils, long, linear chains of β 1-4 linked glucose residues, with the parallel microfibrils being hydrogen bonded. The microfibrils are embedded in other polysaccharides, those of monocotyledonous species are largely composed of arabinoxylans, glucuronoarabinoxylans and mixed-linked β-glucans (Chesson, Gordon and Lomax, 1985) whereas in dicotyledonous species xyloglucans, galactan, arabinogalactans and rhamnogalacturonans (pectin) predominate (Albersheim, 1976). During cellular maturation the secondary cell wall forms inside the primary wall. Being thicker and lignified to varying degrees, the secondary cell wall confers rigidity upon the cell. The secondary cell wall may account for up to 90% of cell wall dry matter (Åman and Graham, 1990), the major constituents being cellulose and xylans.

Degradation of plant polysaccharides

There are two principal determinants of the energy value of a plant polysaccharide: firstly the degree to which it can be hydrolysed to its monomeric constituents and secondly the route by which these monomers are subsequently metabolised.

Starch is the only plant polysaccharide which can be completely hydrolysed by mammalian enzymes. It consists of varying proportions of amylose and amylopectin. Amylose consists of long, linear α 1-4 linked glucose residues, whereas shorter chains which are branched by α 1-6 linkages are found in amylopectin. Starch is digested to dextrins by the α-amylolytic activity of pig pancreatic enzymes in the small intestine. The dextrins are subsequently hydrolysed to glucose by brush-border enzymes prior to absorption. The degree to which starch may be degraded depends on its botanical origin; amylose, for example, is more readily degraded than the more chemically complex amylopectin. Treatments such as pelleting, extrusion and enzyme supplementation, all of which disrupt cell walls to varying extents, result in increased starch digestion in the small intestine of pigs (Graham, Fadel, Newman and Newman, 1989). If the grains are disrupted in some way, pigs can digest cereal starch to a very high degree

(94–99%) (Keys and De Barthe, 1974). However, starch from other sources, e.g., potatoes or legumes are less readily degraded (Graham, 1991) and starch which has been retrograded due to heating and cooling during processing may become resistant to hydrolysis by mammalian α-amylase (Englyst and Cummings, 1985).

Unlike starch, NSP cannot be degraded by mammalian enzymes but it does provide a carbon source for primary fermentation by the gut-microflora. The end-products of such fermentation are acetate, propionate, butyrate, lactate, hydrogen and carbon dioxide with methane being produced as a consequence of further conversion of carbon-dioxide and hydrogen by methanogenic microorganisms. The volatile fatty acids (VFA) usually constitute the major end-products and it is these which are absorbed and subsequently metabolised to yield ATP. Therefore, unlike glucose, the digestion product of starch, the breakdown products of NSP must first be metabolised by the gut-microflora and are not directly available to the animal. Thus the production of ATP from VFA is less efficient than that from glucose, and it is generally accepted that the energy value of carbohydrate fermentation to the pig is about 70% of that of carbohydrate digestion by porcine enzymes (Graham, 1988). Thus, the energy value of a feed depends upon the relative amounts of digestible and fermentable substrates that it contains. The degree to which different types of NSP are fermented depends on various factors including:

- the capacity of microbial enzymes to cleave the glycosidic bonds between carbohydrate and non-carbohydrate moieties
- polymer conformation: long linear molecules which form strong inter-molecular bonds are usually less degradable than highly branched polysaccharides
- in general, the higher the degree of polymerisation the lower the degradability
- the structural association of the polysaccharide with non-carbohydrate components, e.g. associations with lignin will lead to reduced degradability
- physical properties such as water-holding capacity, ion-exchange properties and particle size can all influence NSP degradability

It is clear from the above, that the origin of the NSP must influence the degradability of its constituents: for example at the species level the NSP of soya hulls was highly degraded by growing pigs, in contrast to that of lucerne stems which was poorly degraded (Stanogias and Pearce, 1985). Moreover, at the cellular level, primary cell walls, rich in uronic acids and cellulose may be totally degraded, whereas lignified secondary cell walls where xylose is a major constituent are much less degradable (Chesson, 1990). It is known, however that the ability of the porcine gut-microflora to degrade recalcitrant forms of NSP increases with the age of the pig (Longland, Low, Quelch and Bray, 1993). NSP, particularly when rich in gel-forming constituents such as galacto-mannans, β-glucans or pectins can reduce the absorption of other nutrients from the small intestine leading to reduced nutrient utilisation. Thus, additivity of ingredients in terms of energy evaluation should not always be assumed to be complete.

Measurement of dietary fibre

Traditionally, indirect gravimetric techniques have been used to determine dietary fibre, whereby the insoluble residue remaining after the treatment of a feed with various chemical solutions is said to represent dietary fibre or a specific fraction thereof. However, the crude fibre (CF) method measures mainly cellulose and lignin, but recovery is incomplete. The acid detergent fibre procedure (ADF) developed by van Soest (1963a) is thought to give a reasonable estimate of lignin and cellulose, and although there is recovery of lignin and cellulose by the neutral detergent fibre (NDF) method (van Soest, 1963b), there is some loss of hemicelluloses and particularly pectins, rendering it unsuitable for use with feeds derived from dicotyledonous species which often contain substantial levels of these components.

The above techniques do not measure specific chemical entities and the fractions retained may include material other than NSP and lignin, for example, resistant starch or Maillard products. Analytical techniques have now been developed for routine use which measure directly the individual monomeric constituents of the NSP fraction, the sum of which represents the total NSP content of the sample. Neutral sugars are measured as alditol acetate derivatives of acid hydrolysates of de-starched samples by gas–liquid chromatography, with a separate determination of uronic acids by decarboxylation (Theander and Åman, 1979) or colorimetry (Englyst and Cummings, 1984; Englyst, Quigley, Hudson and Cummings, 1992). Unlike the gravimetric techniques, NSP is measured without loss of major constituent groups of polysaccharides: differences in values obtained for the dietary fibre content of a range of samples by the CF, NDF and NSP techniques are shown in Table 9.1.

For the purposes of trying to evaluate the nutritive value of a feedstuff accurately, it is clear that knowledge of all of its potentially energy-yielding constituents is important. This is well illustrated by the work of Varvaeke, Dierick, Demeyer, Decuypere and Graham (1991) who showed that the

Table 9.1   COMPARISON OF DIETARY FIBRE VALUES FOR FEED SAMPLES DETERMINED BY THREE METHODS (G/KG DM)

Feedstuff	CF	NDF	NSP
Wheat	23	110	110
Barley	52	150	147
Oats	91	248	253
Wheatbran	85	390	360
Maize gluten	74	360	368
Soya hulls	354	599	868
Soyabean meal	85	189	190
Peas	65	109	202
Beans	75	134	170
Rapeseed meal	120	295	286
Lupin	153	258	571
Sugar beet pulp	174	429	635
Citrus pulp	118	321	415
Wheatstraw	400	770	632

hind-gut and faecal digestibility values of crude fibre, ADF and NDF from diets for growing pigs were much lower than the corresponding values for NSP. Thus, use of these gravimetric methods would lead to substantial underestimates of the contribution of NSP to energy balance in these animals.

Anti-nutritive factors (ANF) and their effect on degradation of dietary components

A number of alternative feedstuffs, particularly those regarded as good protein, as well as energy, sources often contain varying amounts of anti-nutritional factors (ANF) which can affect the energy value of a feed. This topic has been recently reviewed by Liener (1990) and thus will only be referred to briefly here. The most common ANF include alkaloids, cyanogenic glycosides, goitrogens, lectins, protease inhibitors and tannins. These ANF may mediate their effects by direct toxicity to the gut-microflora, by inactivating digestive enzymes, by affecting intake or by causing deleterious physiological changes in the pig resulting in reduced utilisation of various nutrients.

Determination of energy value *in vivo*

Gross energy (GE) is the heat of combustion of a unit weight of material (feed, excreta etc.) as determined by bomb calorimetry.

DIGESTIBLE ENERGY (DE)

DE is the GE of a feed minus the GE of the corresponding faeces. The most common method of measuring DE is to collect all faeces produced over a 5–7 day period, following at least a week of adaptation to a change of diet, although there is evidence to suggest that when highly recalcitrant sources of NSP are fed, adaptation periods of 3–5 weeks may be more appropriate (Longland *et al.* 1993). DE values for raw materials are usually derived by difference, since experimental diets almost inevitably contain several energy sources, of which only the test material is unknown. Although additivity of DE values for ingredients is assumed for cereal-based diets this may not always be valid in all circumstances, because inclusion level and type of dietary ingredients, particularly those high in fibre may influence DE (Noblet and Henry, 1993).

Measurement of digestibility overall gives no indication of the site of digestion of the dietary energy. The stomach and small intestine are believed to be the sites where protein, fat and starch are digested by porcine enzymes and where the digestion products are absorbed. By contrast, the fermentation of NSP by the gut-microflora is thought to occur in the large intestine. A number of researchers have therefore placed cannulae in the terminal ileum in order to elucidate sites of digestion/fermentation and thus relative energy yields from different substrates. However, given that some NSP can disappear prior to the terminal ileum (Millard and Chesson, 1984; Graham, Hesselman and man, 1986) it now seems that there is incomplete compartmentalisation of digestion and fermentation within the small and large intestines respectively. Moreover, some starch, perhaps retrograded or encased in the walls of intact cells, may escape undigested from the ileum to be fermented in the hind-gut. Therefore it cannot be certain whether cannulation of the gut can provide values for DE which are much more accurate than those using the total digestibility methods.

Because absorbed hexoses have a higher ATP-generating capacity than VFA, it would seem logical to measure the total amounts of all of these products that pass to the liver. Rérat, Vaugelade and Villiers (1980) described how the rate of blood flow in the hepatic portal vein can be measured, and how with simultaneous determination of arterial and venous nutrient concentrations, absorption can be measured. This approach provides direct information on the partition of carbohydrate, protein and lipid sources between host enzymic activity and microbial fermentation. This method is technically complex, and is clearly inappropriate for routine use, but theoretically at least may be a more useful index of the DE content of diets than the traditional approach.

However, despite the shortcomings of traditional methods of determining DE, the DE system is a commonly used measure of the energy value of pig feeds (ARC, 1981) and has the advantage of being independent of genotype or environment when similar feeding levels are given (Noblet, Le Dividich and Bikawa, 1985).

METABOLIZABLE ENERGY (ME)

ME is DE minus the energy excreted in urine and gases. When cereal-based diets are fed gaseous losses are low ($< 1\%$) and are usually ignored, although up to 3% of GE intake can be lost as gas from diets containing highly fermentable sources of NSP (Zhu, Fowler and Fuller, 1988). Urinary energy depends on the amount of protein in the diet and thus although ME is often regarded as *ca* 5% of total energy excreted, it deviates from this value when there is protein depletion or if excess protein is given to pigs.

NET ENERGY (NE)

NE is ME minus the loss of energy during the digestion of nutrients (the heat increment of feeding). The amount of heat evolved depends on whether the feedstuff is digested by mammalian enzymes or fermented by the gut-microflora, the latter pathway generating more heat. NE represents the energy available to the animal for maintenance and production, and is thus a most useful measure of dietary energy. The classical method of determining NE is that of comparative slaughter. This involves the slaughter of a pre-experimental group of pigs, followed by the measurement of GE intake of the experimental diet by their littermates over a period of at least one month, whereupon the experimental group is slaughtered. The gut contents are removed and all parts of the pigs are recovered, minced and analysed for GE. The difference between the energy content of the pre-experimental pigs and those receiving the test diet is the NE value of the diet. Although this method yields accurate results (Kotarbinska and Kielanowski, 1969), it is a destructive and expensive technique which gives no information on the kinetics of energy use as the animal matures, reproduces or lactates. For kinetic studies, calorimetry, either direct (involving measurement of heat loss) or indirect (involving measurement of inspired and expired gases and the use of well-established equations to determine heat inputs and outputs) is required. Detailed accounts of the use of calorimetry in providing NE values for practical feeding systems have been described by Nehring (1969), Nehring, Schiemann, Hoffman, and Jentsch (1965) and Nehring, Schiemann and Hoffman (1969).

It is clear from the above that *in vivo* determinations of the energy value of a feed are time-consuming, labour intensive and costly in terms of feed, animals and equipment and it would be highly impractical to determine the energy value of every feed or diet in this way. Hence much effort has been expended in seeking to develop rapid, accurate, and cheap laboratory methods for assessing the energy value of pig feeds either by formation of prediction equations based on a) the chemical components of a feed, b) its content of digestible nutrients or c) by the use of *in vitro* digestibility techniques. There have been a number of excellent reviews on the prediction of the energy value of feedstuffs (DE, ME or NE) for pigs (Wiseman and Cole, 1980; Wiseman and Cole, 1983; Noblet and Henry, 1993). Most of the research reported in these reviews was concerned with cereal-based diets, although some of them contained substantial levels of alternative feedstuffs. The following points emerged from this work:

a)   When developing prediction equations, chemical components should be included on the basis of their having large positive or negative effects on digestibility. Fibre has a negative effect on digestibility, as does ash which acts as a diluent. Protein, oil, starch and sugar are positive contributors to dietary energy value.

b)   The accuracy of prediction equations could be increased by developing models for different classes of feedstuff.

c)   The predictive power of equations should be enhanced by use of more detailed chemical analysis of the fibrous fraction to define nutritionally significant entities. Fibre is the greatest chemical source of variation in predicting energy value, accounting for about 70% and 85% of the variation in diets and raw materials respectively (Fernández and Jørgensen, 1986). Thus, it follows that the weakest predictions occur for feedstuffs or diets high in NSP. As digestibility coefficients for total NSP differ widely between feedstuffs (0–100%; Löwgren, Graham, Åman, Raj and Kotarbinska, 1992), it is clear that a single measure of total NSP is insufficient for predicting energy values of alternative feedstuffs. The ADF, NDF and NDF–ADF methods can broadly define cellulose and lignin plus hemicelluloses in graminaceous samples, and NDF has been found to be a better predictor than CF when used for diets containing botanically diverse ingredients (Morgan, Whittemore, Phillips and Crooks, 1987) although it had no advantage over CF within a class of feedstuff (Noblet and Henry, 1993). However, NDF is unsuitable for use with many samples derived from dicotyledonous species, and therefore a more detailed definition of the fibrous fraction by measuring the NSP composition may be required to increase the accuracy of predictions of their energy value.

In view of the above points, the next section of this chapter will explore the effects of including total NSP and its major constituent monomers in addition to NDF as independent variables in data sets for step-wise regression analyses of both individual alternative feedstuffs and diets containing high levels of alternative feeds. Unless otherwise stated, the equations are based on NSP measurements of feedstuffs or diets that were used in energy balance trials using growing pigs at the Institute of Grassland and Environmental Research (IGER), Shinfield from 1985–1991. The chemical constituents which constituted the data set of independent variables were: total NSP, xylose, glucose, uronic acids, NDF, Klason lignin, oil (acid–ether extract – AEE), ash, nitrogen, gross energy, sugars and starch.

Laboratory-based methods for determining the energy value of feedstuffs

As the age of the animal can affect energy digestibility coefficients (Fernández, Jørgensen and Just, 1986) the relationships, prediction equations and *in vitro* methods outlined below are relevant to growing pigs. Unless stated otherwise, DE is expressed as MJ/kg DM and coefficients of energy digestion (DC_e) and metabolism (MC_e) are expressed as a proportion of GE intake. Terms included in equations are as a percentage of DM.

The use of equations based on chemical components

INDIVIDUAL FEEDSTUFFS

There have been few studies assessing the energy value of individual feedstuffs in comparison with those on complete diets. However, if diets are to be formulated to meet certain energy requirements, then knowledge of the energy value of individual ingredients is important. The majority of these studies have used CF, ADF or NDF as measures of the fibrous fraction for inclusion in prediction equations, but there is little information in the literature on the use of NSP or its constituent monomers as predictors of energy value.

The coefficients of energy digestibility (DC_e) and metabolisability (MC_e) of nine alternative feeds which had been determined *in vivo*, were predicted

Table 9.2　CORRELATION COEFFICIENTS (R) BETWEEN CHEMICAL COMPONENTS AND COEFFICIENTS OF ENERGY DIGESTIBILITY (DC_e) AND METABOLISABILITY (MC_e) WHICH HAD BEEN DETERMINED IN VIVO ON INDIVIDUAL ALTERNATIVE FEEDS

	oil	ash	NSP	xylose	glucose	UAC	protein	starch	NDF	lignin	DC_e	MC_e
oil	1.000											
ash	0.550	1.000										
NSP	−0.611	−0.330	1.000									
xylose	0.029	−0.491	0.462	1.000								
glucose	−0.544	−0.650	0.858	0.745	1.000							
UAC	−0.493	0.442	0.322	−0.604	−0.108	1.000						
protein	0.291	−0.142	−0.831	−0.432	−0.639	−0.418	1.000					
starch	0.935	0.328	−0.721	−0.029	−0.589	−0.631	0.569	1.000				
NDF	−0.409	−0.531	0.880	0.811	0.975	−0.411	−0.723	−0.498	1.000			
lignin	−0.271	−0.622	0.534	0.935	0.802	−0.441	−0.444	−0.300	0.821	1.000		
DC_e	−0.041	0.439	−0.478	−0.991	−0.753	0.566	0.486	0.048	−0.820	−0.923	1.000	
MC_e	0.083	0.538	−0.496	−0.966	−0.792	0.545	0.454	0.140	−0.829	−0.953	0.986	1.000

from their measured chemical constituents. Xylose content was found to be inversely correlated with DC_e and MC_e (Table 9.2) and accounted for nearly all of the variance (Table 9.3) (Longland, Low and Bray, unpublished). Xylose has frequently been found to be the least degradable NSP component of a range alternative feeds (Graham *et al.*, 1986; Nordkvist and man, 1986; Longland and Low, 1989), and consequently an inverse relationship between NSP degradability and xylose content is to be expected. This relationship is most likely to influence the DE of feedstuffs in which NSP forms a substantial portion of the DM. The range of alternative feeds was extended to 32, including by-products of graminaceous species, legumes and non-legume dicotyledon species. All had been analysed for the chemical components given above, but for 23, *in vivo* DE values had not been determined; for these samples the mean values of the range cited in the literature were used. When equations to predict the DE of all of the of these feedstuffs were developed, NDF had the highest inverse relationship with DE, and thus was the best single predictor accounting for 63% of the variation and when oil, ash, xylose and uronic acids were included in the equation the R^2 increased to 0.86 (Table 9.3). However when the feeds were grouped and analysed according to botanical source, graminaceous species (n = 13), legumes (high protein) (n = 11) and non-legume dicotyledon species (n = 8), different relationships were obtained. NDF accounted for 88.4% of the variation of the graminaceous species, and the inclusion of ash and starch increased the R^2 to 0.94 (Table 9.3). Both NDF and ash are regarded as negative terms in prediction equations for cereals and thus this type of equation appears to be based on biologically sound principles. For the legumes the most accurate equation (in terms of highest coefficient of determination and lowest coefficient of variation(CV)) was modelled on oil, ash, protein, glucose (from cellulose), and NSP (Table 9.3). As oil, and protein are positive contributors to energy value and ash and NSP have a negative influence (Löwgren *et al.*, 1992), such an equation appears to have

Table 9.3 PREDICTION OF DC_e, MC_e, OR DE BY STEP-WISE REGRESSION ANALYSIS BASED ON THE CHEMICAL COMPOSITION OF INDIVIDUAL FEEDING STUFFS

Intercept slope / regressor	R_2	CV (%)
(1) $DC_e = 0.867 - 0.029_{xylose}$	0.99	3.2
(2) $MC_e = 0.822 - 0.034_{xylose}$	0.97	7.8
(3) $DE = 15.31 - 0.093_{NDF} + 0.269_{oil} - 0.357_{ash} - 0.124_{xylose} + 0.08_{UAC}$	0.86	10.0
(4) $DE = 16.69 - 0.145_{NDF} - 0.289_{ash}$	0.94	8.7
(5) $DE = 12.32 + 0.259_{oil} - 0.618 + 0.111_{protein} + 0.158_{glucose} - 0.1_{NSP}$	0.95	5.6
(6) $DE = 11.576 + 0.052_{starch} - 0.066_{oil} - 0.418_{UAC} + 0.237 - 0.48_{glucose}$	0.99	1.8

Table 9.4   CORRELATION COEFFICIENTS (B) BETWEEN CHEMICAL COMPONENTS AND COEFFICIENTS OF ENERGY DIGESTIBILITY (DC_e) AND METABOLISABILITY (MC_e) OF SEMI-PURIFIED BASAL DIETS CONTAINING VARIOUS LEVELS OF ALTERNATIVE FEEDSTUFFS

	ash	NSP	glucose	xylose	UAC	oil	Starch	sugar	protein	NDF	lignin	DC_e	MC_e
ash	1.000												
NSP	-0.257	1.000											
glucose	-0.516	0.781	1.000										
xylose	0.174	0.311	0.411	1.000									
UAC	0.020	0.498	-0.050	-0.432	1.000								
oil	0.711	-0.368	-0.541	-0.411	0.215	1.000							
starch	-0.500	-0.487	-0.079	-0.164	-0.550	-0.364	1.000						
sugar	0.132	0.095	-0.235	-0.300	0.646	0.283	-0.457	1.000					
protein	0.785	-0.370	-0.565	0.103	-0.013	0.605	-0.300	0.098	1.000				
NDF	0.230	0.175	0.106	0.790	-0.145	-0.135	-0.218	-0.075	0.288	1.000			
lignin	-0.186	0.923	0.870	0.614	0.171	-0.450	-0.414	-0.069	-0.312	0.365	1.000		
DC_e	0.034	-0.604	-0.710	-0.896	0.229	0.449	0.267	0.192	0.123	-0.675	-0.839	1.000	
MC_e	-0.023	-0.564	-0.655	-0.894	0.212	0.389	0.293	0.130	0.065	-0.720	-0.799	0.991	1.000

UAC = uronic acids

biological relevance. For samples derived from non-legume dicotyledon species, the most accurate equation to emerge was based on starch, oil, uronic acids, NSP and glucose (Table 9.3). Oil content was highly positively correlated to xylose content, which has been shown to have a negative influence on DE, and this may account for oil being incorporated as a negative term. Although uronic acids are usually highly digested from this type of sample, they were inversely correlated to starch content, possibly resulting in their inclusion as a negative influence on DE.

Grouping of feeds into different classes improved the predictive power of the equations, and NSP or its constituents were superior to NDF as predictors of DE for alternative, non-graminaceous feeds. However, the DE values of the majority of the feeds represented in equations 3–6 were the mean of a range of published values and therefore should be treated with caution. Had the DE value of these feed samples been determined *in vivo*, then different equations might have emerged. However, the above exercise serves to illustrate that grouping of feedstuffs may result in increased precision of models. As more data becomes available on the NSP content and composition of feeds, together with corresponding *in vivo* measurements of their energy value, then further categorisation might be possible, enhancing the accuracy of prediction.

PREDICTION OF THE ENERGY VALUE OF DIETS

Equations were developed to predict the energy value of 28 semi-purified diets which contained various levels of NSP from nine alternative feedstuffs representing three cereal by-products, two by-products of legumes and four by-products derived from non-legume dicotyledon species. The diets were analysed for total NSP and constituent monomers, nitrogen, ash, oil, starch and sugars, NDF and lignin and the DC_e and MC_e of the diets were determined *in vivo*.

It was found that both xylose and glucose were highly correlated to DC_e and MC_e (Table 9.4).

Table 9.5 PREDICTION OF DC_e AND MC_e BY STEP-WISE REGRESSION BASED ON CHEMICAL CONSTITUENTS OF 28 DIETS

Intercept Slope / Regressor	R_2	*CV (%)*
(7) $DC_e = 0.966 - 0.03485_{xylose}$	0.80	1.4
(8) $DC_e = 0.9806_{xylose} - 0.0283 - 0.00516_{glucose}$	0.94	1.2
(9) $MC_e = 0.944 - 0.03404_{xylose}$	0.79	2.6
(10) $MC_e = 0.955 - 0.02863_{xylose} - 0.00424_{glucos}$	0.90	1.9

Step-wise regression analysis revealed that xylose was the single most potent predictor, accounting for 80% of the variation, and when glucose was added to the equation approximately 92% of the variance was accounted for (Table 9.5).

To test the wider applicability of the above equations to other alternative feedstuffs, equation 8 was applied to 20 semi-purified diets containing various levels of 5 alternative feedstuffs quoted in the literature (Stanogias and Pearce, 1985). The NSP and constituent monomers of diets were calculated from published values. These authors had comprehensively analysed the fibre fraction of the diets in terms of ADF, NDF, cellulose and hemicellulose and thus the choice of appropriate published NSP values for these feedstuffs was facilitated. There was a high ($R^2 = 0.89$) correlation between the predicted DC_e and that determined *in vivo*.

To determine if equation 8 could be applied to more complex diets, the NSP content of 17 diets (4–9 sources of NSP per diet) containing significant but varying amounts of alternative feedstuffs such as sweet potato, rapeseed meal, sunflower meal, peas, maize gluten feed and soyabean meal, the DE, ME and NE of which had been determined *in vivo* (Noblet, Fortune, Dupire and Dubois, 1990) was calculated from information on individual feedstuffs in the literature or for values obtained for similar individual ingredients at IGER. Step-wise regression analysis demonstrated that NDF content accounted for 84% of the variance (CV = 1.8%), the equation describing the relationship being:

$$DC_e = 0.951 - 0.0119_{ndf}$$

The equations which emerged for estimates of ME and NE on the basis of the above chemical constituents were poor predictors of these energy values ($R^2 = < 0.5$).

Given that NSP components (usually xylose and glucose) were the best predictors of the energy value of individual alternative feeds or those incorporated into semi-purified diets, the observation that measured NDF was the most useful single predictor of the DC_e of complex diets was interesting, and may well have reflected the presence of cereal-based ingredients.

However, the use of mean published values to calculate the chemical composition of diets is not ideal, for these values may differ significantly from those of the sample in question. Clearly the more complex the diet, the greater the accumulation of error, and it is possible in the current example that the finally calculated NSP or constituent values bore little relation to their true contents in the diets. In addition to the effects of cultivar, environment, stage of growth etc. on NSP content and composition which may account for some of the differences in NSP content encountered within a feed type, these differences may be enhanced by processing (Table 9.6).

Thus, the NSP content of 'soya bean meal' can range from 170 to 220 g/kg DM depending on the amount of hull remaining in the sample, and the NSP content of soya hulls may range from 620 to 840 g/kg DM depending on the efficiency of the hulling process. Consequently, the problems of accumulating error could be substantial. Therefore, although for reasons of speed and economy the use of tables describing the chemical composition of feedstuffs is attractive, this approach cannot be recommended for alternative feeds, as the range of quoted values may be such that the accuracy of a prediction equation could be severely compromised.

A limitation of using major chemical components on which to base equations to predict the energy value of alternative feeds, is that unless specifically measured, no account will be taken of any effects of the ANF which are characteristic of some alternative feeds. Furthermore, any effects of processing of feedstuffs will not be generally identified by these methods.

NET ENERGY

For the producer, NE may be the most useful system on which to model laboratory methods for predicting the energy value of diets (Hardy, 1991). However, there are no reports in the literature of *in vivo* determinations of the NE of a range of alternative feeds accompanied by full NSP analysis of the diets. Due to the variation in published NSP values for alternative feedstuffs, however, it is doubtful if meaningful equations can be constructed to model the NE value of alternative feeds. It is to be hoped that in the future studies will be undertaken which concomitantly measure the NSP of alternative diets and their NE value. For a comprehensive review on the prediction of the NE value of diets from conventional analyses of

Table 9.6 RANGES OF NON-STARCH POLYSACCHARIDE (NSP) AND CONSTITUENT MONOMER VALUES FOR ALTERNATIVE FEEDSTUFFS (G/KG DM)

Feed	Arabinose	Xylose	Mannose	Galactose	Glucose	Uronic acids	Total NSP
Soya hulls	48–54	67–80	57–64	26–29	312–534	111–129	621–840
Soya meal	16–27	13–24	7–13	34–39	73–83	34–37	168–223
Oat hulls	30–33	286–358	0	9–13	321–381	36–47	691–808
Peas	32–34	19–20	0–3	8–18	66–101	43–58	171–232
Rapeseed meal	50–65	17–26	4–6	18–21	87–107	103–106	279–331
Wheatbran	76–87	110–148	0–3	6–9	101–118	7–12	300–377
Citrus pulp	36–51	15–21	5–11	26–37	109–136	155–210	346–466
Maize gluten	47–91	77–107	1–5	14–19	96–110	12–41	247–373

chemical composition, however, see the recent review by Noblet and Henry (1993).

DIGESTIBLE NUTRIENTS

In this approach, the knowledge that the majority of the variation in the DE value of a diet is accounted for by amounts of digestible fibre, fat, protein and nitrogen-free extract (NFE), is used to predict the energy value of pig feeds. According to Batterham (1990) however, it is probably easier to undertake *in vivo* measurements to determine the energy value of an unknown diet, than it is to determine the digestibility of the fibre, fat, protein and NFE fractions. For practical purposes therefore, the use of tables giving digestibility values for each of the above fractions would be required. These may vary widely between different diets, making it difficult to predict accurately the energy value of an unfamiliar diet.

MEASUREMENT OF ENERGY VALUE *IN VITRO*

A number of *in vitro* assays have been developed to determine the DE value of pig feeds, most of which involve end-point determinations of the energy remaining in feed residues after incubation with commercially available enzymes, porcine digesta or microbial inocula obtained from various portions of the porcine GI tract. Much of the work has concentrated on methods to estimate the energy value of cereals, although some techniques have been modified to include measurements for alternative feeds high in NSP. Thus, methods vary from multienzyme assays attempting to simulate digestion in the foregut or the entire GI tract by the sequential incubation of samples in pepsin, pancreatin and a fibrolytic enzyme preparation (Boisen, 1991) to the use of pepsin followed by incubation in porcine jejunal fluid for DE evaluation of high-starch samples (Furuya, Sakamoto and Takahashi, 1979); for samples containing significant levels of NSP a further incubation in caecal fluid was used, although this latter modification was not very suitable for feeds particularly high in NSP (Varvaeke, Dierick, Decuypere and Hendrickx, 1985). Other methods employ a dialysis membrane to separate products of digestion from undigested residues to overcome any effects of end-product inhibition (Savoie, 1991). Although these methods may achieve reasonable correlations with *in vivo* DE values, they are fairly complex and labour intensive, requiring several incubation steps and various types of enzymes and/or inocula. A slightly different approach has been adopted by Löwgren, Graham and Åman (1989) who developed a method whereby the degradation of diets was determined after incubation with duodenal, ileal or faecal inocula. An interesting feature of this method was that although the rate of degradation of feedstuffs differed between the inocula, there was little difference in the ranking of feeds in terms of energy and NSP degradation. This supports the increasing body of evidence which suggests that fermentation is not confined to the hind-gut. Further studies indicated that the most suitable method for predicting ileal digestibility of starch and crude protein was incubation of the feed in a duodenal inoculum for 12 h, whereas *in vitro* degradation of energy and NSP (and its constituent monomers) from feeds incubated for 48 h in a faecal inoculum, was correlated with the corresponding *in vivo* digestibility coefficients (Graham, Löwgren and Åman, 1989). These authors suggested that by combining data from both duodenal and faecal incubations, it should be possible to estimate both the dietary components which are readily digested and absorbed in the fore-gut and those which are fermented in the hind-gut. These methods are simple one-step incubations, the entire procedure being performed in a single vessel, reducing the errors associated with transfer of residues between enzymes or inocula. A further attraction of the technique is that use of faecal inocula precludes the need for cannulated animals. When 9 alternative feedstuffs (Table 9.7), high in NSP but low in starch were incubated in a faecal inoculum for 48 h (F) there was a high correlation between the DE values determined *in vivo* and energy loss *in vitro* ($R^2 = 0.953$, $p < 0.001$) (equation 11). However, when a further 3 alternative feeds containing intermediate levels of starch (148 to 200 g starch/kg DM) were included in the data set the R^2 was reduced to 0.85, but inclusion of the value obtained after incubation in the duodenal inoculum for 12 h (D) and a measure of energy 'loss' due to fermentation increased this to 0.98 (Table 9.7, equations 12 and 13) (Longland, Low and Bray unpublished).

Equation 11 is based on nine alternative feeds containing substantial amounts of NSP, but little starch. Equations 12 and 13 are based on the addition of a further three NSP-rich alternative feeds which also contained from 150 to 200g starch/kg DM.

In *in vitro* digestion studies it is common to use gravimetric end-point determinations which are destructive. In order to study digestion or fermentation kinetics however, serial harvesting of large numbers of incubation vessels is required. The assumption made is that measurements of residues from a chronological sequence will be continuous, i.e. that digestion in each vessel has occurred at the same rate, but due to slight differences in micro-environments this may not always be the case. Moreover, analysis of serial samples is labour intensive and not suitable as a routine

Table 9.7   PREDICTION OF DE BY STEP-WISE REGRESSION BASED ON IN VITRO INCUBATION OF SAMPLES IN FAECAL OR DUODENAL INOCULA

Intercept slope / regressor	R^2	% CV
(11) DE = −2.27 + 0.171$_F$	0.95	5.8
(12) DE = −1.58 + 0.170$_F$	0.85	8.7
(13) DE = 3.64 − 0.069$_F$ + 0.184$_D$ + 0.113$_{\%ferm}$	0.98	4.3

procedure. However, a highly sensitive non-destructive technique that can distinguish between small differences in fermentation kinetics has been developed for determining the nutritive value of ruminant feedstuffs. This involves using a pressure transducer for measurement of the increase in head-space gas pressure due to the accumulation of gases evolved from the fermentation of samples by rumen microorganisms in gas-tight batch cultures (Theodorou, Williams, Dhanoa, McAllan, and France, 1994). Gas pressure is determined at frequent intervals throughout the incubation, and thus both the rate and extent of fermentation can be determined from each incubation bottle. Furthermore, the effects of ANF on fermentation kinetics can be studied by their introduction to fermentation bottles either at the beginning of the experiment or during established fermentations. Deviations in fermentation kinetics from those of the controls on the one hand or downturns in rates of fermentation on the other can yield valuable information on ANF. This technique has now been automated and requires little labour after the initial preparation of the incubation medium and inoculation of the bottles. This pressure-transducer technique has been adapted to study the fermentation of alternative pig feeds high in NSP by replacing rumen fluid with a porcine faecal inoculum (Longland, Low, Bray and Brooks, unpublished). A significant correlation was found between coefficients of energy digestibility by growing pigs determined *in vivo* and gas production ($R^2 = 0.81$, p < 0.001) for 8 feeds which were by-products of cereals, non-legume dicotyledonous species, and legumes.

The potential for short-term (4 to 5 h) fermentation studies using the manual or automated gas measurement systems is currently being investigated, as this could render the method more suitable for routine use in a commercial environment. Use of a much larger number of samples should establish if this technique can be used as a reliable tool for determining the energy value of pig feeds.

It is clear that *in vitro* methods have considerable potential for reliable prediction of the energy value of feedstuffs for pigs. Providing that sample dilution has not been too great, the effects of ANF which mediate their effects by inhibiting microbial activity or digestive enzymes should be detected. Furthermore, the effects of feed or diet processing should be identified by *in vitro* methods. However, ANF which are detrimental to the pig *per se* or effect intake, transit time or absorption of nutrients will not be detected by these techniques.

IN SACCO METHODS

An alternative approach to estimating the DE value of pig feeds is to introduce nylon bags containing the test feed into the gut, to remove them either *via* cannulae or in the faeces and then chemically analyse the resultant undigested residue. Petry and Handlos (1978) developed a procedure for pigs whereby bags were introduced orally and recovered in the faeces. These workers obtained acceptable relationships between DE values obtained *in sacco* and *in vivo*. Likewise, Graham, Åman, Newman and Newman (1985) introduced bags *via* duodenal cannulae and reported a close correlation ($R^2 = 0.95$) between *in sacco* and *in vivo* organic matter disappearance from 11 samples of feedstuff including cereals and alternative feeds such as distillers grains, rapeseed meal, peas, clover and grass. Other authors have combined *in vitro* and *in sacco* techniques. Thus, Furuya, Sugimoto, Takahashi, and Kameoka (1981) extended their *in vitro* procedure (Furuya *et al.*, 1979) by placing the undigested residue in a nylon bag and introducing it into the terminal ileum, followed by recovery in the faeces. The results of the combined procedure were similar to those obtained *in vivo* for lucerne meal, wheat bran and rice straw. This approach generally requires access to and maintenance of cannulated animals, and unless fitted with re-entrant cannulae rates of digestion of different components are difficult to determine.

Conclusions

From the limited information available, it appears that NSP and some of its constituent monomers (particularly xylose) are superior to NDF as predictors of the energy value of non-graminaceous alternative feeds. Nonetheless, NDF appeared to be a useful predictor of the energy value of samples derived from the *Gramineae*. However, for NSP (or constituents) to be useful predictors of energy value, it is necessary to measure them in each sample, as published values are likely to be unreliable for at least some ingredients. Although a fairly involved technique, NSP analysis can be performed within 48 h, and with modern gas chromatography (GC) and computerised integration systems, this should not pose a problem to commercial companies. However, it is appreciated that this technique may still take too long and be too expensive for use by feed compounders. Grouping of feedstuffs into various classes in terms of botanical origin would seem to enhance the accuracy of prediction equations. Until the time when complete chemical analyses of feedstuffs can be rapidly achieved either by conventional 'wet chemistry', or techniques such as near infrared spectroscopy (NIRS), any effects of ANF and processing will not be accounted for by prediction equations based solely on the major chemical components of feedstuffs. This can be overcome to some degree by the use of *in vitro* procedures. The faecal incubation procedure or the pressure transducer technique show promise for use with alternative feeds high in NSP; neither require cannulated animals, and once established require little attention.

THE WAY FORWARD

For the development of robust, widely applicable equations to predict the energy value of alternative feedstuffs based on chemical constituents, increased data sets of both *in vivo* determinations of energy value (DE, ME or NE) and more detailed chemical analyses (including NSP) of a wide range of alternative feedstuffs, are required. The continued development of *in vitro* techniques which measure fermentability will be of importance for assessing the energy value of feeds high in NSP. Moreover, further development of *in vitro* techniques which require short incubation times would be a major benefit, enabling rapid responses to be made to new batches of feedstuffs soon after their arrival at feed mills.

Acknowledgements

M.S. Dhanoa and J.F. Potter are thanked for their advice on statistics.

References

Albersheim, P. (1976) In *Plant Biochemistry*, pp. 225–272. Edited by J. Bonner and J.E. Varner. New York: Academic Press

Åman, P. and Graham, H. (1990) In *Feedstuff Evaluation,* pp. 161–178 Edited by J.Wiseman and D.J.A. Cole, London: Butterworths

ARC (1981) Commonwealth Agricultural Bureaux, Slough

Batterham, E.S. (1990) In *Feedstuff Evaluation,* pp. 267–282. Edited by J. Wiseman and D.J.A. Cole. London: Butterworths

Boisen, S. (1991) In: in vitro *Digestion for Pigs and Poultry,* pp. 135–145. Edited by M.F. Fuller. Oxford: C.A.B. International

Chesson, A. (1990) In *Feedstuff Evaluation,* pp. 179–196. Edited by J. Wiseman and D.J.A. Cole. London: Butterworths

Chesson, A. , Gordon, A.H. and Lomax, J.A. (1985) *Carbohydrate Research,* 141, 137–147

Englyst, H.N. and Cummings, J.H. (1984) *Analyst,* **109**, 937–942

Englyst, H.N. and Cummings, J.H. (1985) *The American Journal of Clinical Nutrition,* **42**, 778–787

Englyst, H.N., Quigley, M.E., Hudson, G.J. and Cummings, J.H. (1992) *Analyst,* **17**, 1707–1715

Fernández, J.A. and Jørgensen, H. (1986) *Livestock Production Science,* **15**, 53–71

Fernández, J.A., Jørgensen, H. and Just, A. (1986) *Animal Production,* **43**, 127–132

Furuya, S., Sakamoto, K. and Takahashi, S. (1979) *British Journal of Nutrition,* **41**, 511–520

Furuya, S., Sugimoto, N., Takahashi, S. and Kameoka, K. (1981) *Japanese Journal of Zootechnical Science,* **52**, 198–204

Graham H. (1988) *ISI Atlas of Science; Animal and Plant Sciences,* **1**, 76–80

Graham, H. (1991) In: in vitro *Digestion for Pigs and Poultry,* pp. 35–44. Edited by M. Fuller. Oxford: C.A.B. International

Graham, H., Åman, P., Newman, P. and Newman, C.W. (1985) *British Journal of Nutrition,* **54**, 719–726

Graham, H., Fadel, J.G., Newman, C.W., and Newman, R.K. (1989) *Journal of Animal Science,* **67**, 1239–1298

Graham, H., Hesselman, K. and Åman, P. (1986) *Journal of Nutrition,* **116**, 24–251

Graham, H., Löwgren, W. and Åman, P. (1989) *British Journal of Nutrition,* **61**, 689–698

Hardy, B. (1991) In: in vitro *Digestion for Pigs and Poultry,* pp. 181–192. Edited by M.F. Fuller. Oxford: C.A.B. International

Keys, J.E. and De Barthe, J.V. (1974) *Journal of Animal Science,* **39**, 57–62

Kotarbinska, M. and Kielanowski, J. (1969) In *Energy Metabolism of Farm Animals,* pp. 299–301. Edited by K.L. Blaxter, G. Thorbek, and J. Kielanowski. Newcastle-upon-Tyne: Oriel Press

Liener, I.E. (1990) In *Feedstuff Evaluation* pp. 377–394. Edited by J. Wiseman and D.J.A. Cole, London: Butterworths

Longland, A.C. and Low, A.G. (1989) *Animal Feed Science and Technology*, **23**, 67–78

Longland, A.C., Low, A.G., Quelch, D.B. and Bray, S.P. (1993) *British Journal of Nutrition*, **70**, 557–566

Löwgren, W., Graham, H. and Åman, P. (1989) *British Journal of Nutrition*, **61**, 673–683

Löwgren, W., Graham, H., Åman, P., Raj, S. and Kotarbinska, M. (1992) *Animal Feed Science and Technology*, **39**, 183–191

Millard, P. and Chesson, A. (1984) *European Journal of Biochemistry*, **142**, 367–369

Morgan, C.A., Whittemore, C.T, Phillips, P. and Crooks, P. (1987) *Animal Feed Science and Technology* **17**, 81–107

Nehring, K. (1969) In *Energy Metabolism of Farm Animals,* pp. 5–20. Edited by K.L. Blaxter, G. Thorbek and J. Kielanowski. Newcastle-upon-Tyne: Oriel Press

Nehring, K. Schiemann, R., Hoffman, L. and Jentsch, W. (1965) In *Energy Metabolism of Farm Animals,* pp. 243–247. Edited by K.L.Blaxter. London: Academic Press

Nehring, K., Schiemann, R. and Hoffman, L. (1969) In *Energy Metabolism of Farm Animals*, pp. 41–50. Edited by K.L. Blaxter, G. Thorbek, and J. Kielanowski. Newcastle-upon-Tyne: Oriel Press

Noblet, J., Le Dividich, J. and Bikawa, T. (1985) *Journal of Animal Science*, **61**, 452–459

Noblet, J. and Henry, Y. (1993) *Livestock Production Science*, **36**, 121–141

Noblet, J., Fortune, H., Dupire, C. and Dubois, S. (1990) *Journees Recherche Porcine en France*, **22**, 175–184

Nordkvist, E. and Åman, P. (1986) *Journal of the Science of Food and Agriculture*, **37**, 1–7

Petry, H. and Handlos, B.M. (1978) *Archiv für Tierernährung*, **28**, 531–543

Rérat, A., Vaugelade, P. and Villiers, P. (1980) *National Institute for Research in Dairying, Technical Bulletin*, No 3. pp. 177–217. Edited by A.G. Low and I.G Partridge

Savoie, L. (1991) In: in vitro *Digestion for Pigs and Poultry*, pp. 146–161. Edited by M. Fuller. Oxford: C.A.B. International

Stanogias, G. and Pearce, G.R. (1985) *British Journal of Nutrition*, **53**, 513–530

Theander, O. and Åman, P. (1979) *Swedish Journal of Agricultural Research*, **9**, 97–106

Theodorou, M.K., Williams, B.A., Dhanoa, M.S., McAllan, A.B. and France J. (1994) *Animal Feed Science and Technology*, **48**, 185–197

Trowell, H., Southgate, D., Wolever, T., Leeds, A., Gasull, M. and Jenkins, D. (1976) *Lancet*, **1**, 967

van Soest, P.J. (1963a) *Journal of the Association of Official Analytical Chemists*, **46**, 825–828

van Soest, P.J. (1963b) *Journal of the Association of Official Analytical Chemists*, **46**, 829–835

Varvaeke, I.J., Dierick, N.A., Decuypere, J.A. and Hendrickx, H.K. (1985) In *Digestive Physiology of the Pig*, pp. 389–301. Edited by A. Just, H. Jørgensen and J.A. Fernández. Copenhagen: National Institute of Animal Science

Varvaeke, I.J., Dierick, N.A., Demeyer, D.I., Decuypere, J.A. and Graham, H. (1991) *Animal Feed Science and Technology*, **32**, 55–62

Wiseman, J. and Cole, D.J.A. (1980) In *Recent Advances in Animal Nutrition*, pp. 51–67, Edited by W. Haresign. London: Butterworths

Wiseman, J. and Cole, D.J.A. (1983) In *Recent Advances in Animal Nutrition*, pp. 59–70, Edited by W. Haresign. London: Butterworths

Zhu, J-Q., Fowler, V.R. and Fuller, M.F. (1988) *Proceedings of the 4th Symposium on the Digestive Physiology of the Pig*. Edited by L. Burczewska, S. Burczewska, B. Pastuszewska and T. Zebrowska. Jablonna, Poland: Polish Academy of Sciences, Institute of Animal Physiology and Nutrition.

10

CHOICE-FEEDING SYSTEMS FOR PIGS

S.P. ROSE[1] AND M.F. FULLER[2]
[1] *Harper Adams College, Newport, Shropshire TF10 8NB, UK*
[2] *The Rowett Research Institute, Greenburn Road, Bucksburn, Aberdeen AB2 9SB, UK*

Introduction

There has been a recent revival of interest in the use of choice-feeding systems in which pigs are given the freedom to select from more than one feed. The interest stems from the proposition that pigs recognise both their own nutritional needs and the properties of food which satisfy those needs. This is called nutritional wisdom. If choice-fed pigs have sufficient nutritional wisdom then individual animals could select a mixed diet that accurately meets their requirements. These individuals could also make daily changes to their nutrient intakes without having to rely solely on changes in their voluntary intakes of a single conventional diet. The system could potentially increase the efficiency of feed utilization in practical pig production systems.

The objectives of this paper are:

- to examine whether choice-fed pigs are able to control, closely and consistently, their nutrient intakes
- to discuss how a choice-feeding regimen could be used in practical production systems
- to examine published comparisons of choice-feeding and conventional single feed systems.

How much nutritional wisdom do pigs have?

The extent of the nutritional wisdom possessed by pigs is still poorly understood. The average diet selection of a pen of pigs is difficult to interpret because individuals within the pen could select very differently.

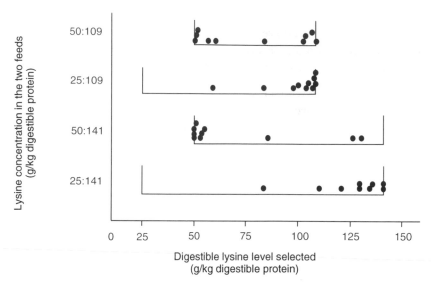

Figure 10.1 Lysine concentrations selected by individual pigs offered various dietary choices. Each dot represents the lysine concentrations selected by individual pigs that were offered two feeds that varied only in lysine concentration. Four treatment combinations were compared. For each treatment, the two vertical lines indicate the lysine concentration of the two feeds offered. The horizontal line that connects the two vertical lines indicates the range of lysine concentrations that could be selected by an individual pig.

The pen mean may be merely an average that may not represent the choice of any individual. Data collected on individually housed pigs are more valuable in examining the extent of their abilities to select diets.

CAN PIGS DISCRIMINATE BETWEEN FEEDS WITH DIFFERENT NUTRIENT COMPOSITIONS?

There is convincing evidence that pigs are able to distinguish between two feeds which differ in protein or amino acid concentration. There is also evidence that they prefer to eat a feed that is adequate in these nutrients than one that has a large excess. This has been demonstrated by work on selection for lysine (Fairley, Rose and Fuller, 1993).

Figure 10.1 shows the results of an experiment in which individual pigs were given a choice of two feeds that differed only in lysine concentration. Four different combinations of two feeds were offered to the pigs after an initial 'training period'. The 'training period' ensured that the pigs gained experience of both feeds by giving them access to each on alternate days.

The same four feeds, and additional blends of the high- and low-lysine feed, were also given as complete single feeds to growing pigs. The data from this second trial showed that there was a growth response to increasing lysine concentration, but there was no significant further increase in growth rate after the lysine concentration reached 46 g/kg digestible protein.

A number of conclusions can be made from these data.

1. No pig chose a diet with a lysine concentration below 46 g/kg digestible protein. All therefore preferred feeds that avoided a lysine deficiency.
2. The pigs avoided a very high lysine diet if there was a large difference in lysine concentration between the two feeds and if the lower lysine feed was not lysine-deficient. A lysine concentration closest to, but not less than, 46 g/kg digestible protein was preferred.
3. The clustering of points towards the ends of each range show that pigs tended to prefer one diet or the other, depending on the particular combination offered. Their preferences therefore appear to be controlled by the nutritional effects of the diets rather than by the taste preferences of the pigs or an aversion to added lysine.

Experiments that used similar protocols have been completed with threonine (Fairley, Rose and Fuller, 1993) and tryptophan (Fairley, Rose and Fuller, 1995). The results lead to the same general conclusion that pigs can certainly discriminate between diets which differ only in the concentration of a single amino acid.

Kyriazakis, Emmans and Whittemore (1990) offered individual pigs a choice of two feeds that differed in crude protein concentration. As with amino acid choices, the pigs preferred the feeds that were closest to the crude protein requirement of the group, which was estimated, by feeding the pigs the diets singly, as being between 174 to 213 g/kg. However, two pigs out of the 24 in the choice experiment preferred a feed that had significantly less protein than 174 g/kg.

CAN PIGS SELECT DIFFERENT PROPORTIONS OF TWO FEEDS TO MEET THEIR REQUIREMENTS?

Pigs appear to have little or no ability to make a daily selection of different proportions of two feeds in order to meet their nutrient requirements. The lysine choice-feeding data in Figure 10.1 can again be used as an example. Three quarters of the 40 choice-fed pigs in the experiment selected mostly (> 0.85) one feed. The pigs' feed preferences seem to have been dominated by the avoidance of lysine deficiency. One quarter of the pigs ate a mixture of the two feeds (> 0.15 of each), but only a few pigs selected a lysine concentration that was likely to be optimal for their growth (according to current recommendations). Fairley (personal communication) observed that the number of pigs which ate a mixture of the two feeds was reduced if the 'training period' was extended from 8 to 14 days. Perhaps the pigs that ate both feeds after the shorter training period were demonstrating an inability to detect the nutritional characteristics of one or other feed rather than a precise nutritional wisdom.

Gill (1994) gave individual pigs a choice between a feed with high protein but low lysine concentration and a feed with low protein but high lysine concentration. Although most pigs ate only the low-protein, high-lysine feed, he studied the feeding behaviour of the smaller number of pigs that chose a mixture of the two feeds. He concluded that these pigs had no consistent pattern of eating the two alternative feeds: the less-preferred feed was eaten infrequently and apparently randomly. There was no evidence that, by their feeding activity, the pigs were attempting to regulate amino acid or protein intake or absorption from the digestive tract.

Pigs have a similar lack of ability (or desire) to mix two feeds that differ in threonine and tryptophan concentrations (Fairley *et al.*, 1993, 1994). A similar experiment Henry (1993) with older pigs given the choice between two feeds differing in lysine concentration showed that the pigs chose a mixture of the two feeds. However, the protocol did not include a 'training period'. In addition, the feeder positions were exchanged weekly. It is likely that there was a significant period at the beginning of each week when the pigs needed to sample both feeds before re-establishing a preference. The selections of individual pigs were not given, but the coefficients of variation of the weekly means were large. Many of the pigs may have been selecting predominantly one feed towards the end of each week.

Pigs given a choice of feeds varying in crude protein content have the same feeding behaviour as was seen with single amino acids (Kyriazakis *et al.*, 1990). Almost two thirds of the 24 pigs given a choice between high- and low-protein feeds selected predominantly (> 0.85) one feed (Kyriazakis, 1989). The preferred feed was generally the one closest to the requirement. However, with one particular combination, a very high-protein feed and a very low-protein feed, pigs tended on average to choose a mixture of the two, although their selections were highly variable and there was no evidence that any pig consistently chose a mixture of the two appropriate to its needs.

In summary, the pig has only a rudimentary ability to select an appropriate diet when given a choice of two feeds. Pigs have aversions to feeds which have serious deficiencies or excesses of some nutrients. A high proportion of choice-fed pigs select predominantly one feed and there is no evidence that the majority of pigs can select different proportions of two feeds to achieve their optimum daily nutrient intakes.

Choice-feeding systems

Even though pigs have an imprecise ability to choose an ideal diet it is still possible that a practical production system could use a choice-feeding system. This section examines the possibilities for implementing choice-feeding in pig growing and finishing systems. Only the parameters of productive efficiency are considered although there may also be welfare considerations.

EQUIPMENT NEEDS

A choice-feeding system imposes extra costs in feeding equipment. Two feed storage bins rather than one would usually be needed and two feed troughs would be needed in each pen even though the pen size might only justify one. A sophisticated feed delivery system would be needed to ensure that the correct feed was dropped into the correct feed hopper.

These extra feeding costs are small relative to the total cost of feed in a pig growing and finishing unit. However, a choice-feeding system would also need additional management. Choice-feeding has to lead to a significant improvement in production efficiency to be cost-effective and to persuade pig farmers to adopt it.

WHAT SELECTION TO OFFER

Energy and protein are the two most costly components of a practical pig feed. A choice-feeding system that would allow growing or finishing pigs to choose an appropriate energy:protein ratio would therefore have most practical benefit. It is difficult to supply to all pigs at all times the optimum ratio of energy:protein in a practical diet. The optimum ratio changes continuously as pigs grow and is also affected by such environmental factors as ambient temperature. The genetic potential and sex of the pigs can also affect the energy:protein ratio they require.

A choice of two protein concentrations

A choice of two feeds that have the same energy concentration but which differ in protein concentration would allow pigs to select between two

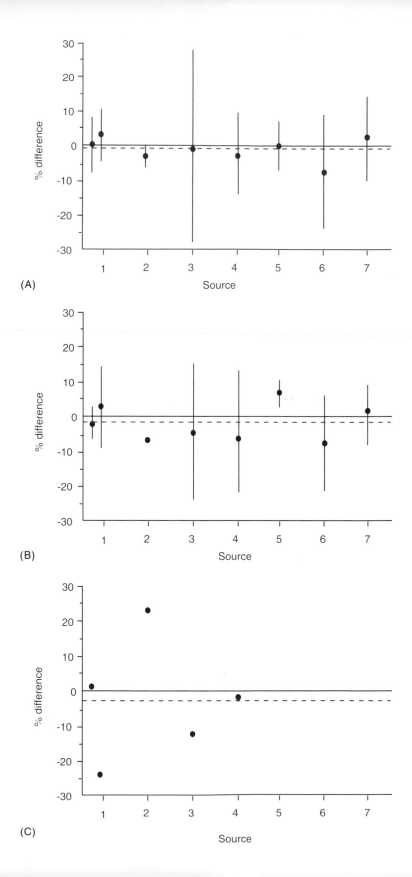

(A)

(B)

(C)

Figure 10.2 (opposite) A) Body weight gain; B) Gain:feed; C) Weight gain (g)/g protein
A summary of seven experiments in which the performance of growing pigs given a choice of high- and low-protein feeds was compared with that of pigs given a single complete diet. Each value is the percentage difference between the choice-fed pigs and the pigs given the single diet, with 95% confidence intervals where available. Growth rates, feed conversion efficiencies and protein conversion efficiencies that were better or worse than those of pigs given complete single feeds are shown above or below the 0% line respectively. The dotted line shows the mean difference for all experiments. Sources: 1. Gill, Sanchez-Serrano, English, Robledo and Roden (1994), grower and finisher stages shown separately; 2. Dalby, Forbes, Varley and Jagger (1994); 3 and 4. Bradford and Gous (1991b); 5. Bradford and Gous (1991a); 6. Bradford and Gous (1992); 7. Kyriazakis, Emmans and Taylor (1993).

energy:protein ratios. Both feeds would contain the correct concentrations of all other nutrients required for growth.

Seven experiments in the literature were examined in which this type of choice-feeding system has been compared directly with a conventional single feed system. Choice-feeding experiments were only considered if the range of protein concentrations offered spanned the protein concentration of the single feed. The results are summarized in Figure 10.2. The weight gains and the ratios of gain:feed and gain:protein of the choice-fed pigs were expressed as a percentage of the values from the pigs given the single feed. Growth rates, feed conversation efficiencies and protein conversion efficiencies that were better or worse than those of pigs given complete single feeds are shown above or below the 0% line respectively.

The mean growth rate and feed conversion efficiency of the choice-fed pigs were 1.06% and 1.78% lower respectively ($P > 0.05$) than those of the pigs given the single diet. The choice-fed pigs had a mean protein conversion efficiency (kg of weight gained:per kg of protein consumed) 2.5% worse ($P > 0.05$) than that of the pigs given the single diet but results were very variable. This no doubt depends on the protein concentration of the single diet against which choice feeding is compared. If this is substantially higher than the pigs' requirement, choice-fed pigs have the opportunity to maintain performance on a diet of lower protein concentration.

In summary, there is no evidence that there is any improvement in growth or efficiency in pigs given a choice between a high- and a low-protein feed. The increased capital cost of implementing a choice-feeding system is therefore unlikely to be justified by any improvement in pig performance.

Table 10.1   THE PRODUCTIVE PERFORMANCE AND ECONOMIC
COMPARISON OF GROWING PIGS GIVEN EITHER A COMPLETE SINGLE FEED OR
TWO CHOICE-FEEDING REGIMENS FOR 12 WEEKS (Rose, 1994, unpublished)

Treatments	Live weight gain (kg/pig)	Feed intake (kg/pig)	Proportion of wheat selected (kg/kg)	Feed conversion ratio (kg/kg)	Feed cost (£/pig)	Feed cost per kg live weight gain (pence/pig)
Complete single feed	69.0	155.9		2.266	28.68	41.6
Choice-feeding treatments						
Balancer: ground wheat	61.4	161.8	0.472	2.635	26.79	43.6
Balancer: whole grain wheat	55.8	160.8	0.621	2.879	25.75	46.1
SEM (9 df)	2.33	9.11	0.087	0.1137		

A choice between a cereal and a balancer feed

Cereals usually account for more than 50% of the cost of a complete pig
feed. They provide digestible energy at lower cost than most other feedstuffs
and they also contribute a significant amount of protein (albeit not well-
balanced protein) to the diet. Choice-feeding regimens have been examined in which pigs have been given a choice of two feeds based on a
complete balanced diet, the cereal component and a balancer feed that
consists of all the other components of the balanced diet. The balancer feed
generally contains protein concentrates, minerals, essential fatty acids,
vitamins and perhaps also some cereal. This combination of feeds allows
the pigs to select the energy:protein ratio of their diet. The vitamin and
mineral concentrations in the balancer feed can be adjusted to provide
optimum intakes of these nutrients.

Work at Harper Adams College has examined a choice-feeding system
using wheat as the cereal (Rose, unpublished data). A conventional pig
grower feed was formulated that contained, per tonne, 723 kg wheat, 50 kg
barley, 109.5 kg soya bean meal, 30 kg meat and bone meal, 75 kg fish meal
and 12.5 kg of vitamin and mineral supplement. A balancer feed was
formulated that had 446 kg wheat, 100 kg barley, 219 kg soya bean meal,
60 kg meat and bone meal, 150 kg fish meal and 25 kg of vitamin and
mineral supplement per tonne. Forty-eight pigs (mean liveweight 18.5 kg)
were given one of three treatments for twelve weeks. They received either
the single complete grower feed alone or a choice of the balancer feed and
wheat. The wheat was given either ground or as whole grains. Although
whole grain wheat is poorly digested it was thought that the reduced milling

and transport costs might outweigh the nutritional disadvantages. The results are shown in Table 10.1.

The choice-fed pigs had a 15% lower growth rate ($P < 0.01$) and an 18% poorer feed conversion efficiency ($P < 0.01$) than the pigs given the complete single feed (Table 10. 1). The choice-fed pigs that were offered the balancer feed and ground wheat selected an overall diet that was similar in composition to the complete single feed. However, the weekly selections of wheat and the selections between individual pens were highly variable. This might account for the lower mean growth rate of the choice-fed group. The choice-fed pigs offered whole grain wheat selected greater ($P < 0.05$) proportions of wheat but did not change their total feed intakes. The pigs given this treatment tended ($P < 0.1$) to have lower live weight gains and poorer feed conversion efficiencies. Choice-feeding reduced the overall feed costs but the feed cost per kg of live weight gain was greater than with the complete single feed.

Few other studies have examined the use of wheat in choice-feeding systems, but there has been an interest in the use of maize that has spanned four decades. A quantitative review of the productive performance of these choice-fed pigs versus pigs given a complete single feed considered twelve published experiments. The same procedures as were described earlier were used to compare the data. The results are summarised in Figure 10.3.

Choice-fed pigs had mean growth rates and feed conversion efficiencies that were poorer by 6.6% and 0.4% respectively than those of single fed pigs. Although these differences would be commercially important, they were not statistically significantly different ($P > 0.05$).

In summary, choice-feeding systems that offer a cereal with a balancer feed have the potential to reduce feed costs. However, growth rates and feed conversion efficiencies tend to be reduced ($P < 0.1$) and so the cost of producing each kg of live weight gain may be greater.

Conclusions

Although pigs can detect differences between feeds in nutrient composition, certainly as regards amino acid composition, they do not select different proportions of two feeds to meet accurately and consistently their daily requirements. Practical choice-feeding systems for pigs incur extra costs in equipment and management time. There is no evidence that a system of providing a choice between high- and low-protein feeds gives any advantage in productive performance. A choice-feeding system that uses a

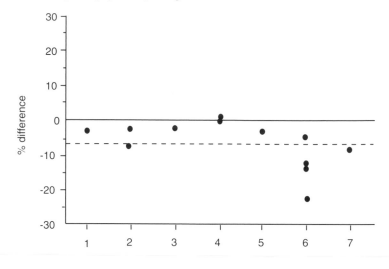

(A)   Source

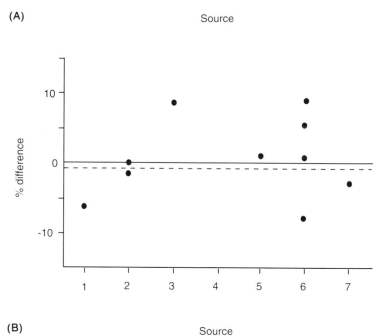

(B)   Source

Figure 10.3 A) Body weight gain; B) Gain: feed A summary of 12 experiments in which the performance of pigs given a choice of maize and a balancer feed was compared with that of pigs given a single complete feed. Each value is the mean percentage difference between the choice-fed pigs and those given the single feed. Growth and feed conversion efficiencies of choice-fed pigs that were better or worse than those of pigs given complete single feeds are shown above and below the 0% line respectively. The dotted line shows the mean difference for all experiments. Sources: 1. Bradford and Gous (1991b); 2. Ekstrom and Mahan (1976); 3. Foster, Jones and Pickett (1964); 4. Holck and Tribble (1965); 5. Rérat and Henry (1964); 6. Hutchinson, Terrill, Jensen, Becker and Norton (1957); 7. Overfield, Cromwell and Hays (1968).

balancer feed and a cereal may reduce feed costs but weight gains and feed conversion efficiencies are poorer.

Braude (1965) reviewed the use of complete single feeds and choice-feeding and concluded that the latter is doomed when very high and efficient performance is aimed at. The evidence produced since that time does not seem to justify any change to that opinion.

Acknowledgements

This work was supported in part by the Scottish Office Agriculture and Fisheries Department.

References

Bradford, M.M.V. and Gous, R.M. (1991a) *Animal Production*, **52**, 185–192

Bradford, M.M.V. and Gous, R.M. (1991b) *Animal Production*, **52**, 323–330

Bradford, M.M.V. and Gous, R.M. (1992) *Animal Production*, **55**, 227–232

Braude, R. (1965) *Proceedings of the Nutrition Society*, **26**, 163–181

Dalby, J.A., Forbes, J.M., Varley, M.A. and Jagger, S. (1994) *Animal Production*, **58**, 435–436

Ekstrom, K.E. and Mahan, D.C. (1976) *Ohio Swine Day Reports*, 56–61

Fairley, R.A.C., Rose, S.P. and Fuller, M.F. (1992) *1st Joint Anglo/French Symposium on Human and Animal Nutrition, Rennes.* p. 192

Fairley, R.A.C., Rose, S.P. and Fuller, M.F. (1993) *Animal Production*, **56**, 468–469

Fairley, R.A.C., Rose, S.P. and Fuller, M.F. (1995) *Proceedings of the Nutrition Society*, (in press)

Foster, J.R., Jones, H.W. and Pickett, R.A. (1964) *Purdue University Research Progress Report*, **No 103**, pp. 1–5

Gill, J.B.R. (1994) *BSc Investigational Thesis*, Harper Adams College

Gill, B.P., Sanchez-Serrano, A.P., English, P.R., Robledo, M. and Roden, J. (1994) *Animal Production*, **58**, 435

Henry, Y. (1993) *Reproduction, Nutrition, Development*, **33**, 489–502

Holck, G.L. and Tribble, L.F. (1965) *Journal of Animal Science*, **24**, 887

Hutchinson, H.D., Terrill, S.W., Jensen, A.H., Becker, D.E. and Norton, H.W. (1957) *Journal of Animal Science*, **16**, 562–567

Kyriazakis, I. (1989) PhD. Thesis, University of Edinburgh

Kyriazakis, I., Emmans, G.C. and Whittemore, C.T. (1990) *Animal Production*, **51**, 189–199

Kyriazakis, I., Emmans, G.C. and Taylor, A.J. (1993) *Animal Production*, **56**, 151–154

Overfield, J.R., Cromwell, G.L. and Hays, V.W. (1968) *Progress Report of the Kentucky Agricultural Experimental Station*, pp. 64–65

Rérat, A. and Henry, Y. (1964) *Annales de Biologie animale, Biochimie et Biophysique*, **4**, 441–444

11

THE INFLUENCE OF VITAMINS ON REPRODUCTION IN PIGS

B.P. CHEW

Department of Animal Sciences, Washington State University, Pullman, WA 99164-6320, USA

The importance of vitamins in reproduction has long been recognized. Dietary deficiencies of certain vitamins will decrease reproductive efficiency by directly or indirectly impairing ovarian function and the uterine environment, thereby influencing oestrus, conception and pregnancy. This chapter discusses the influence of the antioxidant vitamins, vitamin A, β-carotene, vitamin E and vitamin C, together with the B-complex vitamin folic acid, on reproduction in pigs. This does not imply that other vitamins are not necessary for this important function. Where necessary, relevant studies in other animals will also used.

Vitamin A

Vitamin A exists in three naturally-occurring forms: retinol, retinoic acid and retinal (Frickel, 1984). These retinoids differ only in their chemical oxidation state and each form plays a somewhat unique role in regulating physiological functions. For example, retinoic acid is unable to support certain aspects of male and female reproduction or vision (Zile and Cullum, 1983).

Preformed vitamin A is found mainly in the animal kingdom while plants do not contain significant amounts of vitamin A. Vitamin A is essential for the maintenance of reproductive function and foetal development (Thompson, Howell and Pitt, 1964). The most characteristic symptoms of vitamin A deficiency in sows include the birth of weak, dead or malformed pigs (Palludan, 1975). Vitamin A could exert these effects by influencing ovarian steroidogenesis and the uterine environment.

Rats reared on a vitamin A-deficient diet have a reduced ability to secrete progesterone and 20a-hydroxypregn-4-en-3-one into the ovarian venous blood on days 9 and 15 of pregnancy (Ganguly, Pope, Thompson, Toothill, Edwards-Webb and Waynforth., 1971; Ganguly and Waynforth, 1971). Collagenase-dispersed porcine luteal cells showed increased production of progesterone when incubated in the presence of retinol and retinoic acid (Talavera and Chew, 1988). The stimulatory effects of retinol and retinoic acid on progesterone production have also been shown with granulosa cells from rats (Bagavandoss and Midgley, 1987). The effect of retinoic acid on ovarian cell steroidogenesis is somewhat unexpected because vitamin A-deficient, retinoic acid-fed female rats showed decreased ability to secrete progesterone (Ganguly *et al.*, 1971; Ganguly and Waynforth, 1971) and increased foetal resorption, the latter due to a generalized necrosis of the functional zone of the placenta. The observed difference may have been due to an insufficient concentration of retinoic acid reaching the reproductive organ in retinoic acid-supplemented animals or due to an artifact of *in vitro* experiments. However, the importance of retinoic acid for male and female reproduction has been demonstrated and will be discussed later.

Besides its reported effects on ovarian steroidogenesis, vitamin A could directly influence the uterine environment and the development of the embryo and foetus or it could indirectly affect the uterine environment by affecting ovarian progesterone production. Vitamin A-deficiency in pregnant pigs produced structural and compositional changes in placental glycosaminoglycan (Steele and Froseth, 1980). Also, the pig uterus secretes a large amount of several proteins in response to progesterone (Roberts and Bazer, 1980, 1988). These uterine proteins are very important to the nutrition of the conceptus (Roberts and Bazer, 1980; Buhi, Baler, Ducsay, Chun and Roberts, 1979). This is especially true in the pig because the porcine trophoblast does not invade the uterine epithelium but rather remains in superficial attachment to the uterine surface. For instance, uteroferrin, an iron-containing purple glycoprotein (Roberts and Bazer, 1980) secreted by the pig uterine endometrium, is responsible for transporting iron to the conceptus (Buhi *et al.*, 1979). This suggests the possible existence of other transport proteins in uterine secretions. Indeed, retinol-binding protein (RBP) and retinoic acid-binding protein (RABP) have been found in endometrium, ovary, testis, and other tissues of mammalian species (Ong and Chytil, 1975; Chytil, Page and Ong, 1975). Because vitamin A is essential for the maintenance of reproductive function and foetal development (Thompson *et al.*, 1964), Adams, Bazer and Roberts (1981) therefore attempted to identify possible vitamin A-carrier proteins that could transport vitamin A from the maternal uterine endometrium to the conceptus. They reported the presence of RBP in uterine secretions from pigs in the luteal phase of the oestrus cycle. Furthermore, the uterine secretion of RBP was progesterone-induced, as assessed by hormone replacement therapy using ovariectomized gilts (Adams *et al.*, 1981). The uterine RBP identified (Adams *et al.*, 1981) was unique to the uterus in that it differs from serum and cellular RBP in its binding affinity for retinol. These findings were supported by a subsequent study (Clawitter, Trout, Burke, Araghi and Roberts, 1990) using pseudopregnant and progesterone-treated ovariectomized pigs. These authors reported that porcine uterine RBP is made up of at least four distinct proteins which share some sequence homology with the NH_2-terminal of human serum RBP; the most acidic isoform shows the highest (70%) while the most basic polypeptide shows the least (30%) sequence identity.

Besides the pig uterine endometrium, the pig conceptus also produces RBP (Harney, Mirando, Smith and Bazar., 1990; Trout, Kramer, Tindle, Farlin, Baumbach and Roberts, 1990). Harney *et al.* (1990) demonstrated that RBP secretion by pig conceptus occurs throughout the peri-implantation period, with secretion occurring as early as day 10 of pregnancy, prior to the onset of conceptus elongation. The conceptus RBP shares very high amino acid sequence homology with human and rabbit serum RBP (Harney *et al.*, 1990). Therefore, pig conceptus RBP is more homologous with human serum RBP than RBP secreted into the uterine lumen of pigs. The gene encoding the RBP secreted by the early pig conceptus was recently cloned (Trout *et al.*, 1990). These researchers isolated two apparently full length cDNA clones (approximately 900 base pairs) for day 13 to 17 porcine conceptus RBP. They demonstrated that porcine conceptus RBP and human serum RBP share 90% nucleic acid sequence identity within the coding region and 37% identity within the 3' non-coding region (Trout *et al.*, 1990).

MECHANISM OF ACTION

Retinoids are lipid-soluble. Therefore, it is not surprising to find retinoids associated with water-soluble retinoid-binding proteins in plasma (Kanai *et al.*, 1968), the cytoplasm (Bashor, Foft and Chytil, 1973; Chytil and Ong, 1984), and nucleus (Petkovich, Brand, Kurst and Chambon, 1987; Giguere, Ong, Segui and Evans, 1987) of cells. Obviously, a similar situation holds true in the case of the uterine milieu. Adams *et al.* (1981) reported that total vitamin A in uterine secretions increased in progesterone-treated pigs, thereby suggesting an increased transport of nutrients across the uterine epithelium during early pregnancy when blood progesterone is elevated. This observation coupled with the increased secretion of RBP by both the uterine endometrium and the conceptus during the peri-implantation period suggest increased local transport of retinoids by RBP to the developing conceptus. The mechanism for the cellular uptake of retinoids is not yet fully elucidated (Blomhoff, Green, Berg and Norum, 1990). Retinoids could be taken up by target cells through a receptor-mediated mechanism, through non-specific spontaneous transfer or through fluid-phase endocytosis (Blomhoff *et al.*, 1990). Upon entering the cell, the retinoid is then bound to cellular retinoid-binding proteins. Cellular binding-proteins for retinol (CRBP), retinoic acid (CRABP) and reti-naldehyde (CRALBP) have all been identified (Blomhoff *et al.*, 1990). In addition, the presence of nuclear retinoic acid receptors was recently reported (Petkovich *et al.*, 1987; Giguere *et al.*, 1987; Giguere, Shago, Zirngibl, Tate, Rossant and Varmula, 1990; Research News, 1990). At least three distinct forms of nuclear retinoic acid receptors (RARα, RARβ and RARγ) have been identified. Another distinctly different nuclear acid retinoic acid receptor (RXRα) has recently been cloned (Mangelsdorf, Ong, Dyck and Evans, 1990). These receptors belong to the superfamily of hormone receptors that bind to DNA.

Because the cellular retinoid-binding proteins are primarily found in the cytosolic compartment and with the recent discovery and characterization of nuclear retinoic acid receptors, two hypotheses for the role of cellular retinoid-binding proteins have been suggested. First, these cytoplasmic carrier proteins could serve to transport bound retinoid to sites of metabo-lism. An example of this is the cellular RBP type II (Ong, Kakkad and McDonald, 1987) which makes bound retinol available for esterification by membrane-bound enzymes of the small intestine. Secondly, these cytoplas-mic proteins could serve to transport bound retinoids to the nucleus, where the retinoids (retinoic acid), but not the binding protein, become bound to the nuclear retinoic acid receptor. These cytoplasmic carrier proteins (especially cellular retinoic acid-binding protein) play an important role in the movement of retinoids to the nucleus, thereby regulating the concen-tration of the retinoids. However, it is the nuclear retinoic acid receptors which function as direct agents in gene regulation related to cell differen-tiation. Recent evidence indicates that the mechanism of action for retinoic acid is similar to those described for steroid hormones, thyroid hormones and vitamin D$_3$ (Petkovich *et al.*, 1987; Giguere *et al.*, 1987; Dolle, Ruberte, Kastner, Petkouich, Stoner, Gudas and Chambon, 1989).

What then is the role of retinoids in conceptus development? Retinoids may directly affect embryo development by regulating cell differentiation and proliferation (Schindler, 1986), transcription of specific genes (Chiocca,

Davis and Stein, 1988, 1989; Bedo, Santisteban and Aranda, 1989) or possibly act as a morphogen (Thaller and Eichele, 1987; Research News, 1990). Alternatively, retinoids may indirectly influence embryo development by regulating ovarian steroid production (Talavera and Chew, 1988), immune cell function (Chew, 1987), and interferon production (Blalock and Gifford, 1976, 1977), all of which may affect the establishment and maintenance of pregnancy. Through these combined actions, retinoids may regulate early embryonic elongation, placentation and subsequent conceptus development.

β-Carotene

Carotenoids are found widely distributed in nature where they play an important role in protecting the functional integrity of cells. Research on the role of carotenoids in animal nutrition has, in the past, been hampered by the assumption that β-carotene merely serves a provitamin A role. However, studies have shown that carotenoids play an important role in regulating reproduction function in animals.

In contrast to vitamin A, studies on β-carotene and reproduction are comparatively few. A major reason for this is the previous assumption that β-carotene merely serves a provitamin A role. Suggestions for a direct role of β-carotene in regulating reproductive processes have been provided in cattle (Lotthammer., 1979; Rakes, Owens, Britt and Whitlow, 1985; Snyder and Stuart, 1981). The group at Washington State University (Michal, Chew, Wong, Heirman and Standaert, 1990) reported a lower incidence of retained placenta in dairy cows supplemented with β-carotene than in unsupplemented cows or cows fed vitamin A. Retained placenta often leads to metritis and decreased reproductive performance (Chew, Keller, Erb and Malven, 1977). In contrast, others (Wang, Larson, and Owens, 1982; Wang, Hafi and Larson, 1988; Folman *et al.*, 1979) have failed to report a beneficial effect of β-carotene on reproduction. This discrepancy may be due to differences in the initial β-carotene or vitamin A status of the experimental animals (Graves-Hoagland, Hoagland and Woody, 1989).

Rats injected with β-carotene showed higher foetal growth rate and lower pup mortality than control animals (Chew and Archer, 1983). Similarly, cross-bred gilts injected weekly with 228 mg β-carotene starting on the day of breeding and continuing through weaning at 3 weeks postpartum had lower embryonic mortality, larger litter size, and heavier litter weight at birth and at weaning than unsupplemented gilts (Brief and Chew,

1985). This is in general agreement with a later study (Coffey and Britt, 1993) in which multiparous sows were injected once at weaning with 0, 50, 100 or 200 mg β-carotene. The authors reported a linear increase in litter size at birth with increasing dosage of β-carotene injected. Also, the number of piglets born dead was lower in treated pigs. In contrast, no treatment effect was observed with first litter gilts treated similarly. It is unclear from the latter study whether increased litter size at birth was due to increased ovulation rate or decreased embryonic mortality. Because uterine secretions are very important to the survival and development of the embryo and because these uterine secretions are progesterone-induced (Bazer, 1975), it is possible that the lower embryonic mortality observed (Brief and Chew, 1985; Coffey *et al.*, 1993) may be due to changes in the uterine secretions and ovarian progesterone production. Indeed, gilts injected every other day, starting on the day of breeding, with 16.4 mg β-carotene had higher quantities of uterine-specific proteins on day 15 of pregnancy than did untreated pigs (Chew, Rasmussen, Pubols and Preston, 1982). Also, β-carotene stimulated progesterone production by collagenase-dispersed pig luteal cells ten fold after a 24 h incubation period (Talavera and Chew, 1988). The magnitude of response observed with β-carotene was several fold higher than that observed with retinol or retinoic acid. *In vitro*, β-carotene also stimulated progesterone production by bovine luteal cells; again, the stimulation profile for β-carotene was different from that observed with the retinoids (Talavera and Chew, 1987). Further evidence for the specific role of β-carotene in regulating reproduction was provided by O'Fallon and Chew (1984) who showed that β-carotene is an integral component of the microsomal membrane of bovine luteal cells. Recently, it was reported (Chew, Szenci, Wong, Gilliam, Hoppe and Coelho, 1994a,b) that β-carotene, when injected intramuscularly into mature pigs, was taken up in significant amounts by ovarian follicles and corpora lutea. In the corpora lutea, the β-carotene was found in the nuclei, mitochondria and microsomes, suggesting possible direct or indirect action in these subcellular organelles.

MECHANISM OF CAROTENOID ACTION

β-carotene possesses potent antioxidant properties. Reactive oxygen species are generated in biological systems. These reactive species disrupt cell membrane function and induce DNA single-strand breaks. Free radicals react with polyunsaturated fatty acids, nucleotides in DNA and proteins in cells, causing tissue damage. Therefore, β-carotene (and other carotenoids) may enhance reproductive functions by protecting the highly active ovarian follicles/corpora lutea and the uterine cells from free radical damage, thereby protecting important cell surface receptors and intracellular organelles. This will allow for optimal steroidogenic activity of ovarian cells and optimal secretion of critical uterine proteins.

Carotenoids also may regulate cell function by regulating a number of other separate or interrelated cell events (Chew, 1993). In an attempt to study the possible intracellular regulation by β-carotene, the uptake of β-carotene by the bovine corpus luteum was studied (O'Fallon and Chew, 1984). It was shown that β-carotene is mainly associated with the nuclei and the mitochondria. In addition, it was found that β-carotene is an integral component of the microsomal membrane of bovine luteal cells. These data showed the presence of β-carotene in cellular organelles and suggested that β-carotene may have a biological function at these sites.

β-carotene supplementation seems to improve reproductive performance in pigs by influencing ovarian steroidogenesis, by altering the uterine mileu, or by some yet unknown mechanism. If β-carotene plays a direct role, then supplementation is important because the basal pig diet is generally low or devoid of β-carotene. The availability and the route of administration of β-carotene will need to be addressed because little or none is absorbed by the pig.

Vitamin E

Vitamin E is a term used to describe 8 naturally-occurring fat-soluble nutrients called tocopherols. α-tocopherol is the most biologically active and the most abundant isomer of the vitamin E family.

A number of studies have reported a beneficial effect of supplemental vitamin E in improving litter size in pigs (Adamston, Krider and James, 1949; Malm, Pond, Walker, Homan, Aydin and Kirkland, 1976; Chavez and Patton, 1986; Mahan, 1991, 1994). Mahan (1991) fed gilts 0.3 mg/kg Se plus 0, 16, 33 or 66 IU DL-α-tocopherol acetate/kg diet over 3 parities. Litter size at birth was only increased in gilts supplemented with vitamin E (10.87, 11.20, 10.04 piglets/litter, respectively) compared with control (9.85 piglets/litter). However, the incidence of agalactia was higher in gilts supplemented with lower amounts of vitamin E. Consequently, the number of live piglets on day 7 was higher as the level of vitamin E supplementation increased (7.67, 8.43, 9.04, 9.28 piglets/litter for 0, 16, 33 and 66 IU vitamin E, respectively). A more recent study (Mahan, 1994) suggested that 44 to 66 IU vitamin E/kg diet approached the optimum level for vitamin E supplementation to pigs for maximum litter size. Concentrations of vitamin E in the colostrum and milk of these vitamin E supplemented pigs also were increased, suggesting possible health benefits to the suckling piglets. Presently, the NRC (1988) recommends 22 IU vitamin E/kg diet for breeding swine.

MECHANISM OF ACTION

α-tocopherol is the principal lipid-soluble chain-breaking antioxidant in plasma and tissues (Jialal and Grundy, 1992). It functions by trapping peroxyl free radicals. In mammalian cells, α-tocopherol is mainly located in mitochondria fractions and in endoplasmic reticulum, whereas little is found in cytosol and peroxisomes (Bjorneboe, Bjorneboe and Drevon, 1990). A low molecular weight α-tocopherol-binding protein has been identified in the cytosol of rodent liver and heart (Dutta-Roy, Gordon, Leishman, Paterson, Duthie and James, 1993). This binding protein is believed to be responsible for the intracellular transport of α-tocopherol.

Vitamin E reacts with peroxyl radicals produced from polyunsaturated fatty acids in membrane phospholipids or lipoproteins to yield a stable lipid hydroperoxide.

Through this biochemical reaction, vitamin E effectively reduces harmful lipid free radicals and thereby protects tissues from free radical attack. In its antioxidant function, vitamin E acts in concert with other enzyme systems, notably glutathione peroxidase, to protect cells. Selenium is an important component of this metalloenzyme. While vitamin E scavenges radicals within the membrane phase, glutathione peroxidase blocks the initiation of lipid peroxidation from within the soluble phase of the cell. Therefore, vitamin E and selenium act in synchrony to prevent cell lipid peroxidation from oxygen radicals which are produced in reproductive organs like the corpus luteum during regression. In fact, ovarian content of vitamin E (and other antioxidant vitamins) changes dramatically during development and functional regression of the corpus luteum (Aten, Duarte and Behrman, 1992). Vitamin E and selenium may guard against peroxidation of arachidonic acid (Lawrence, Mathias, Nockels and Tengerdy, 1985; Likoff, Cuptill, Lawrence, McKay, Mathias, Nockels and Tengerdy, 1981) and thereby alter arachidonic acid metabolism in animals (Aziz and Klesius, 1986; Bryant and Bailey, 1980). This mechanism could account for the observed enhanced reproductive response by maintaining the functional integrity of reproductive organs.

Vitamin C

Vitamin C or L-ascorbic acid is the most important antioxidant in extracellular fluid (Stocker and Frei, 1991). Research on the possible benefits of supplementing vitamin C to pigs on reproduction has not yielded promising results. Yen and Pond (1983) reported no improvement of reproduction or lactation performance of pigs supplemented with 1 g ascorbic acid/day during gestation. However, vitamin C is important for the health and survival of animals (Chew, 1995). Due to the instability of ascorbic acid, dietary supplementation may need to be much higher than 1 g/day to elicit a response, at least on reproduction.

MECHANISM OF ACTION

Vitamin C can protect biomembranes against lipid peroxidation damage by eliminating peroxyl radicals in the aqueous phase before these can initiate peroxidation (Frei *et al.*, 1989). Vitamin C is effective against superoxide, hydroxyl radical, hydrogen peroxide, peroxyl radical and singlet oxygen (Sies, Stahl and Undquist, 1992). Vitamin C also may function to reduce the tocopheroxyl radical, thereby restoring the radical scavenging activity of vitamin E (Niki, 1987). The ascorbate radical (semi-dehydroascorbate) is reduced to ascorbate by NADH-dependent semi-dehydroascorbate reductase (Green and O'Brien, 1973). The sparing effect of high dietary vitamin C in elevating plasma and tissue vitamin E has been reported in guinea pigs (Bendich, D'Apolito, Gabriel and Machlin., 1984).

Folic acid

Ensminger, Colby and Cunma (1951) published the first study on the importance of folic acid on reproduction in pigs. However, the majority of subsequent studies in this area were conducted in the last 10 years. In an early report (Otel, Costin and Oeriu., 1972), sows injected with 750 mg folycysteine (folic acid chemically bound to cysteine) on days 1 and 9 postmating farrowed larger litters (10.00 vs 8.34 piglets/litter) than uninjected sows. Subsequent studies with injectable folic acid showed similar improvement in litter size in pigs. Matte, Girard and Brisson (1984) reported a significant increase in the total number of piglets born and number of piglets born alive in sows injected intramuscularly every 2 weeks with 15 mg folic acid throughout gestation. Furthermore, they demonstrated that, when compared with non-injected control pigs, the magnitude of increase in the number of live piglets born in folic acid injected sows was greater for flushed sows (12.0 vs 10.9 piglets/litter) than for sows fed a normal intake (10.7 vs 10.5 piglets/litter). Increase in the prolificacy of sows given parenteral administration of folic acid has similarly been reported by others (Friendship and Wilson, 1991).

The possible beneficial effect of dietary folic acid supplementation on reproduction in the pig also has been investigated. Tremblay, Matte, Dufour and Brisson (1989) fed sows 5 mg folic acid/kg of diet when fed a normal intake, flushed or stimulated with pregnant mare's serum gonadotrophin (PMSG). Folic acid supplemented sows had higher foetal survival rate and lower early embryonic/foetal death during the first 30 days of gestation. The reduction in early embryonic death was more pronounced in flushed and PMSG-stimulated sows. The lower embryonic death in folic acid fed sows may be explained through changes in uterine secretion of prostaglandins. Sows supplemented with 15 mg folic acid/kg diet from the time of mating had a 3-fold higher total uterine PGF2α and PGE2 (Laforest, Farmer, Girard and Matte, 1993). These prostaglandins are important to embryo implantation and survival. Lindemann and Kornegay (1989) similarly reported that sows fed folic acid at 1 mg/kg during gestation and lactation farrowed 0.38, 0.65 and 1.8 more piglets than unsupplemented sows for parities 1, 2 and 3, respectively.

In contrast to these reports which consistently showed increased litter size in pigs supplemented with folic acid, especially in multiparous sows, a recent large multiparity study involving five research stations (Harper, Lindemann, Chiba, Combs, Hanblin, Korneghy and Sourthern, 1994) failed to substantiate these findings. In this study, folic acid was fed at 1, 2 and 4 mg/kg and had no effect on the total number of piglets born, or on the number of live piglets at birth and at day 21, or on piglet weight. The authors suggested differences in body stores and folic acid in feedstuffs as possible explanations for this discrepancy. Matte, Girard and Tremblay (1993) fed gilts 5 or 15 mg folic acid from 9 weeks of age through to 7 weeks of gestation and failed to observe an improvement in age at puberty, litter size or litter weight. The authors suggested that folic acid is beneficial only under conditions of high reproductive capacity, for example, in multiparous and flushed/superovulated females.

The NRC (1988) recommendation for folacin in breeding pigs is 0.3 mg/kg. This value is based on a typical corn-soybean meal diet which is estimated to contribute 0.3 mg/kg.

MECHANISM OF ACTION

Folic acid, a B-complex vitamin, is involved in the transfer and utilization of monocarbon units, in the synthesis of RNA, DNA and neurotransmitters, and in the interconversion of amino acids. As such, the demand for folic acid is expected to increase during periods of rapid tissue growth. Indeed, the above evidence supports the need for supplemental folic acid during gestation, a period involving the growth and differentiation of embryonic and placental tissues, and of increased secretion of uterine proteins critical embryonic survival.

References

Adams, K.L., Bazer, F.W. and Roberts. R.M. (1981) Progesterone-induced secretion of a retinol-binding protein in the pig uterus. *Journal of Reproduction and Fertility,* **62**:39

Adamstone, F.B., Krider, J.L and James, M.F. (1949) Response of swine to vitamin E-deficient rations. *Annals of the New York Academy of Sciences,* **52**:260

Aten, R.F., Duart, K.M. and Behrman, H.R. (1992) Regulation of ovarian antioxidant vitamins, reduced glutathione, and lipid peroxidation by luteinizing hormone and prostaglandin F2α. *Biology of Reproduction,* **46**:401

Aziz, E. and Klesius, P.H. (1986) Effect of selenium deficiency on caprine polymorphonuclear leukocyte production of leukotriene B4 and its neutrophil chemotactic activity. *American Journal of Veterinary Research,* **47**:426

Bagavandoss, P. and Midgley, Jr. A.R. (1987) Lack of difference between retinoic acid and retinol in stimulating progesterone production by luteinizing granulosa cells *in vitro. Endocrinology,* 121:420

Bashor, M.M., Toft, D.O. and Chytil, F. (1973) *In vitro* binding of retinol to rat tissue components. *Proceedings of the National Academy of Sciences, USA,* **70**:3483

Bazer, F.W. (1975) Uterine protein secretions: relationship to development of the conceptus. *Journal of Animal Science,* **41**:1376.

Bedo, G., Santisteban, P. and Aranda, A. (1989) Retinoic acid regulates growth hormone gene expression. *Nature,* **339**:231

Bendich, A., D'apolito, P., Gabriel E. and Machlin, L.J. (1984) Interaction of dietary vitamin C and vitamin E on guinea pig immune responses to mitogens. *Journal of Nutrition,* **114**:1588

Bjorneboe, A., Bjorneboe, G.E. and Drevon, C.A. (1990) Absorption, transport and distribution of vitamin E. *Journal of Nutrition,* **120**:233

Blalock, J.E. and Gifford, G. (1976) Suppression of interferon production by vitamin A. *Journal of General Virology*, **32**:143

Blalock, J.E., and Gifford, G. (1977) Retinoic acid (vitamin A acid)-induced transcriptional control of interferon production. *Proceedings of the National Academy of Sciences, USA*, **74**:5382

Blomhoff, R., Green, M.H., Berg, T. and Norum, K.R. (1990) Transport and storage of vitamin A. *Science*, **250**:399

Brief, S. and Chew, B.P. (1985) Effects of vitamin A and β-carotene on reproductive performance in gilts. *Journal of Animal Science*, **60**:998

Bryant, R.W. and Bailey, J.M. (1980) Altered lipoxygenase metabolism and decreased glutathione peroxidase activity in platelets from selenium-deficient rats. *Biochemical and Biophysical Research Communication*, **92**:268

Buhi, W., Bazer, F.W., Ducsay, C., Chun, P.W. and Roberts, R.M. (1979) Iron content, molecular weight and possible function of the progesterone-induced purple glycoprotein. *Federal Proceedings*, **38**:733

Chavez, E.R. and Patton, K.L. (1986) Response to injectable Se and vitamin E on reproductive performance of sows receiving a standart diet. *Canadian Journal of Animal Science.* **66**:1065

Chew, B.P. (1987) Vitamin A and β-carotene on host defense. *Journal of Dairy Science*, **70**:232

Chew, B.P. (1993) Effects of supplemental β-carotene and vitamin A on reproduction in swine. Symposium on 'High supplemental vitamin levels and reproductive function of swine'. *Journal of Animal Science*,**71**:247

Chew, B.P. (1995) Antioxidant vitamins affect food animal immunity and health. Symposium on 'Beyond deficiency: New views on vitamins in ruminant nutrition and health' *Journal of Nutrition*, **125**:1804S

Chew, B.P. and Archer, R.G. (1983) Comparative role of Vitamin A and β-carotene on reproduction and neonate survival in rats. *Theriogenology*, **20**:459

Chew, B.P., Keller, H.F., Erb, R.E. and Malven, P.V. (1977) Peripartum concentrations of prolactin, progesterone and the estrogens in the blood plasma of cows retaining and not retaining fetal membranes. *Journal of Animal Science*,**44**:1060

Chew, B.P., Rasmussen, H., Pubols, M.H. and Preston, R.L (1982) Effects of vitamin A and β-carotene on plasma progesterone and uterine protein secretion in gilts. *Theriogenology*, **18**:643

Chew, B.P., Szenci, O., Wong, T.S., Gilliam, V.L., Hoppe, P.P. and Coelho, M.B. (1994a) Uptake of β-carotene by plasma, follicular fluid, granulosa cells, luteal cells and endometrium in pigs after administrating injectable β–carotene. *Journal of Animal Science*, **77**:100

Chew, B.P., Szenci, O., Wong, T.S., Gilliam, V.L., Hoppe, P.P. and

Coelho, M.B. (1994b) Subcellular distribution of β-carotene in ovarian luteal cells of pigs injected with β-carotene. *Journal of Animal Science,* **77**:100

Chiocca, E.A., Davies, P.J.A. and Stein, J.P. (1988) The molecular basis of retinoic acid action. *Journal of Biological Chemistry,* **263**:11584

Chiocca, E.A., Davies, P.J.A. and Stein. J.P. (1989) Regulation of tissue transglutaminase gene expression as a molecular model for retinoid effects on proliferation and differentiation. *Journal of Cellular Biochemistry,* **39**:293

Chytil, F., Page, D.L. and Ong. D.E. (1975) Presence of cellular retinol and retinoic acid binding proteins in human uterus. *International Journal of Vitamin Research,* **45**:293

Chytil, F., and Ong. D.E. (1984) Cellular retinoid-binding proteins. In: M. B. Sporn, A. B. Roberts, D. S. Goodman (ed.) *The Retinoids,* vol. 2. p. 89. Academic Press, New York

Clawitter, J., Trout, W.E., Burke, M.G., Araghi, S. and Roberts, R.M. (1990) A novel family of progesterone-induced, retinol-binding proteins from uterine secretions of the pig. *Journal of Biological Chemistry,* **265**:3248

Coffey, M.T. and Britt, J.H (1993) Enhancement of sow reproductive performance by β-carotene or vitamin A. *Journal of Animal Science,* **71**:1198

Dolle, P., Ruberte, E., Kastner, P., Petkovich, M., Stoner, C.M., Gudas, L.J. and Chambon, P. (1989) Differential expression of genes encoding α, β and γ retinoic acid receptors and CRABP in the developing limbs of the mouse. *Nature,* **342**:702

Dutta-Roy, A.K., Gordon, M.J., Leishman, D.J., Paterson, B.J., Duthie G.G. and James, W.P.T. (1993) Purification and partial characterisation of an α-tocopherol-binding protein from rabbit heart cytosol. *Molecular and Cellular Biochemistry,* **123**:139

Ensminger, M.E., Colby, R.W. and Cunha. T.J. (1951) Effect of certain B-complex vitamins on gestation and lactation in swine. *Station Circular: Washington Agricultural Experimental Station,* no. 134

Folman, Y., Ascarelli, I., Herz, Z., Rosenberg, M., Davidson, M. and Halevi, A. (1979) Fertility of dairy heifers given a commercial diet free of β-carotene. *British Journal of Nutrition,* **41**:353

Frei, B., England, L. and Ames, B.N. (1989) Ascorbate is an outstanding antioxidant in human blood plasma. *Proceedings of the National Academy of Sciences, USA.* **86**:6377

Frickel, F. (1984) Chemistry and physical properties of retinoids. In: M.B. Sporn, A.B. Roberts, D.S. Goodman (ed.) *The Retinoids,* vol. 1, p. 7, Academic Press, New York

Friendship, R.M. and Wilson, M.R. (1991) *Canadian Journal of Veterinary Research,* **32**:565

Ganguly, J., Pope, G.S., Thompson, S.Y., Toothill, J., Edwards-Webb, J.D. and Waynforth, H.B. (1971) Studies on the metabolism of vitamin A: The effect of vitamin A status on the secretion rates of some steroids into the ovarian venous blood of pregnant rats. *Biochemistry Journal,* **122**:235

Ganguly, J. and Waynforth, H.B. (1971) Studies on the metabolism of vitamin A: The effect of vitamin A status on the content of some steroids in the ovaries of pregnant rats. *Biochemistry Journal,* **123**:669

Giguere, V., Ong, E.S., Segui, P. and Evans, R.M. (1987) Identification of a receptor for the morphogen retinoic acid. *Nature,* **330**:624

Giguere, V., Shago, M., Zirngibl, R., Tate, P., Rossant, J. and Varmuza, S. (1990) Identification of a novel isoform of the retinoic acid receptor (gamma) expressed in the mouse embryo. *Molecular and Cellular Biology,* **10**:2335

Graves-Hoagland, R.L., Hoagland, T.A. and Woody, C.O. (1989) Relationship of plasma β-carotene and vitamin A to luteal function in postpartum cattle. *Journal of Dairy Science,* **72**:1854

Green, R.C and O'Brien, P.J. (1973) The involvement of semidehydro-ascorbate reductase in the oxidation of NADH bylipid peroxide in mitochondria and microsomes. *Biochimica et Biophysica ACTA.* **293**:334

Harney, J.P., Mirando, M.A., Smith, L.C. and Bazer, F.W. (1990) Retinol-binding protein: A major secretory product of the pig conceptus. *Biology of Reproduction,* **42**:523

Harper, A.F., Lindemann, M.D., Chiba, L.I., Combs, G.E., Handlin, D.L., Kornegay, E.T. and Southern, L.L. (1994) An assessment of dietary folic acid levels during gestation and lactation on reproductive and lactational performance of sows: a cooperative study. *Journal of Animal Science,* **72**:2338

Jialal, I. and Grundy. S.M. (1992) Influence of antioxidant vitamins on LDL oxidation. *Annals of the New York Academy of Sciences,* **669**:237

Kanai, M., Raz, A. and Goodman, D.S. (1968) Retinol-binding protein: The transport protein for vitamin A in human plasma. *Journal of Clinical Investigation*

Laforest, J.P., Farmer, C, Girard, C.L. and Matte, J.J. (1993) *Proceedings of the 4th International Conference on Pig Reproduction, Univ. Missouri, Columbia,* MO.

Lawrence, L.M., Mathias, M.M., Nockels, C.F. and Tengerdy, R.P. (1985) The effect of vitamin E on prostaglandin levels in the immune organs of chicks during the course of an E. coli infection. *Nutrition Research,* **5**:497

Likoff, R.O., Guptill, D.R., Lawrence, L.M., McKay, C.C., Mathias,

M.M., Nockels, C.F. and Tengerdy, R.P. (1981) Vitamin E and aspirin depress prostaglandins in protection of chickens against *Escherichia coli* infection. *American Journal of Clinical Nutrition,* **34**:245

Lindemann, M.D., and Kornegay, E.T. (1989) Folic acid supplementation to diets of gestating-lactating swine over multiple parities. *Journal of Animal Science,* **67**:459

Lotthammer, K. H. (1979) Importance of β-carotene for the fertility of dairy cattle. *Feedstuffs,* **51**:6

Mahan, D.C. (1991) Assessment of the influence of dietary vitamin E on sows and offspring in three parities: reproductive performance, tissue tocopherol, and effects on progeny. *Journal of Animal Science,* **69**:2904

Mahan, D.C. (1994) Importance of the antioxidant vitamins in swine nutrition – applications of principles. *55th Minnesota Nutrition Conference,* p. 43

Malm, A., Pond, W.G., Walker, Jr. E.F., Homan, M., Aydin, A. and Kirtland, D. (1976) Effect of polyunsaturated fatty acids and vitamin E level of the sow gestation diet on reproductive performance and on level of alpha tocopherol in colostrum, milk and dam and progeny blood serum. *Journal of Animal Science,* **42**:393

Mangelsdorf, D.J., Ong, E.S., Dyck, J.A. and Evans, R.M. (1990) Nuclear receptor that identifies a novel retinoic acid response pathway. *Nature,* **345**:224

Matte, J.J., Girard, C.L. and Brisson, G.J. (1984) Folic acid and reproductive performances of sows. *Journal of Animal Science,* **59**:1020

Matte, J.J., Girard, C.L. and Tremblay, G.F. (1993) Effect of long-term addition of folic acid on folate status, growth performance, puberty attainment, and reproductive capacity of gilts. *Journal of Animal Science,* **71**:151

Michal, J.J., Chew, B.P., Wong, T.S., Heirman, L.R. and Standaert, F.E. (1990) Effects of supplemental β-carotene on blood and mammary phagocyte function in peripartum dairy cows. *Journal of Dairy Science,* **73** (Suppl. 1):149 (Abstr.)

Niki, E. (1987) Interaction of ascorbate and α-tocopherol. *Annals of the New York Academy of Sciences,* **498**:186

NRC. (1988) Nutrient requirements of swine. (9th Ed.). National Academy Press. Washington, D.C.

O'Fallon, J.V., and Chew, B.P. (1984) The subcellular distribution of β-carotene in bovine corpus luteum. *Proceedings of the Society for Experimental Biology and Medicine,* **177**:406

Ong, D.E., and Chytil, F. (1975) Retinoic acid-binding protein in rat tissue. *Journal of Biological Chemistry,* **250**:6113

Ong, D.E., Kakkad, B. and McDonald, P.N. (1987) Acyl-CoA-independent esterification of retinol bound to cellular retinol-binding protein (Type II) by microsomes from rat small intestine. *Journal of Biological Chemistry,* **262**:2729

Otel, V., Costin, Gh. and Oeriu, I. (1972) The use of folcysteine for the control of embryo mortality in pigs. II. Results of large-scale experiments. *Journal of Veterinary Medicine, Series A,* **19**:766

Palludan, B. (1975) II. The influence of vitamin A on reproduction in sows. Danish–USSR Symp. Vitamins & Trace Minerals in Anim. Nutr., Moscow

Petkovich, M., Brand, N.J., Krust, A. and Chambon, P. (1987) A human retinoic acid receptor which belongs to the family of nuclear receptors. *Nature,* **330**:440

Rakes, A.H., Owens, M.P., Britt, J.H. and Whitlow, L.W. (1985) Effects of adding β-carotene to rations of lactating cows consuming different forages. *Journal of Dairy Science,* **68**:1732

Research News (1990) The embryo takes its vitamins. *Science,* **250**:372

Roberts, R.M., and Bazer, F.W. (1980) The properties, function and control of synthesis of uteroferrin, the purple protein of the pig uterus. In: M. Beato (ed.) *Steroid Induced Proteins,* Elsevier-North Holland, Amsterdam

Roberts, R.M., and Bazer, F.W. (1988) The functions of uterine secretions. *Journal of Reproduction and Fertility,* **82**:875

Schindler, J. (1986) Retinoids, polyamines, and differentiation. In: M.I. Sherman (ed.) *Retinoids and Cell Differentiation,* p. 137. CRC Press, Boca Raton

Sies, H., Stahl W. and Sundquist, A.R. (1992) Antioxidant functions of vitamins: vitamins E and C, β-carotene, and other carotenoids. *Annals of the New York Academy of Sciences,* **669**:7

Snyder, W.E., and Stuart, R.L. (1981) Nutritional role of beta-carotene in bovine fertility. *Journal of Dairy Science,* **64** (Suppl. 1):104 (Abstr.)

Steele, V.S., and Froseth, J.A. (1980) Effect of gestational age on the biochemical composition of porcine placental glycosaminoglycans. *Proceedings of the Society for Experimental Biology and Medicine,* **165**:480

Stocker, R. and Frei, B. (1991) Oxidants and Antioxidants. In: *Oxidative Stress* (Sies, H., ed.) pp. 213. Academic Press, London

Talavera, F., and Chew, B.P. (1987) *In vitro* interaction of lipoproteins with retinol, retinoic acid and β-carotene on progesterone secretion by bovine luteal cells. *Journal of Dairy Science,* **70** (Suppl. 1):225 (Abstr.)

Talavera, F. and Chew, B.P. (1988) Comparative role of retinol, retinoic acid and β-carotene on progesterone secretion by pig corpus luteum *in vitro. Journal of Reproduction and Fertility*, **82**:611

Thaller, C. and Eichele, G. (1987) Identification and spatial distribution of retinoids in the developing chick limb bud. *Nature*, **327**:625

Thompson, J.N., Howell, J.M. and Pitt, G.A.J. (1964) Vitamin A and reproduction in rats. *Proceedings of the Royal Society of Biology*, **195**:510

Tremblay, G.F., Matte, J.J., Dufour, J.J. and Brisson, G.J. (1989) Survival rate and development of fetuses during the first 30 days of gestation after folic acid addition to a swine diet. *Journal of Animal Science*, **67**:724

Trout, W.E., Kramer, K.K., Tindle, N.A., Farlin, C.E., Baumbach, G.A. and Roberts, R.M. (1990) Molecular cloning of retinol binding protein secreted by early pig conceptuses. *Biology of Reproduction*, **42** (Suppl. 1):167 (Abstr.)

Wang, J.Y., Hafi, C.B. and Larson. L.L. (1988) Effect of supplemental β-carotene on luteinizing hormone released in response to gonadotropin-releasing hormone challenge in ovariectomized holstein cows. *Journal of Dairy Science*, **71**:498

Wang, J.Y., Larson, L.L. and Owen, F.G. (1982) Effect of beta-carotene supplementation on reproductive performance of dairy heifers. *Theriogenology* **18**:461

Yen, J.T. and Pond, W.G. (1983) The response of swine to periparturient vitamin C supplementation. *Journal of Animal Science*, **56**:621–624

Zile, M.H., and Cullum, M.E. (1983) The function of vitamin A: Current concepts. *Proceedings of the Society for Experimental Biology and Medicine*, **172**:138

12

AMINO ACID REQUIREMENTS OF BREEDING PIGS

J.E. PETTIGREW
Department of Animal Science, University of Minnesota, St. Paul, Minnesota, 55108 USA

Introduction

. This paper addresses quantitative amino acid requirements of gestating and lactating sows. Requirements of breeding boars, of developing gilts, and of sows between weaning and rebreeding are important, but are beyond the scope of this paper.

A method for estimation of the quantitative requirements is proposed. The method, which is described more comprehensively elsewhere (Pettigrew, 1993), uses a combination of empirical and factorial approaches. It has many shortcomings because of the inadequacy of the data available, but is suggested to have significant advantages over other methods proposed to date. It should be used only until a more reliable method becomes available. Dourmad, Etienne and Noblet (1991) estimated amino acid requirements of sows using an approach that shares some characteristics with the one described here.

Maintenance requirements

There is continuous unavoidable sloughing of cells from tissues such as skin and intestinal mucosa. The rate of activity of amino acid degrading metabolic pathways may never reach zero. These represent obligatory losses of amino acids from the body, which must be replaced in order to maintain constant conditions. The amount needed for this purpose is usually estimated as an intercept in a regression equation, and is called a maintenance requirement.

The most recent and most reliable estimates of amino acid needs for

Table 12.1 ESTIMATED OBLIGATORY LOSSES (MAINTENANCE REQUIREMENTS) FOR AMINO ACIDS[a]

Amino Acid	MG/KG $W^{0.75}$
Arginine	0
Histidine	0
Isoleucine	20
Leucine	29
Lysine	49
Methionine	11
Total sulphur amino acids	61
Phenylalanine	23
Total aromatic amino acids	46
Threonine	41
Tryptophan	14
Valine	25

[a] Values, except threonine, from Fuller *et al.* (1989) adjusted for digestibility. Threonine value deduced as described in text.

maintenance of pigs are those of Fuller, McWilliam, Wang and Giles (1989). Those estimates are used herein, except for threonine. The maintenance requirement for threonine was estimated by a process described below. The value obtained is considerably lower than that of Fuller *et al.* (1989), but similar to other published estimates.

These maintanence requirements were estimated using diets containing a combination of crystalline amino acids and casein, a very digestible protein. Therefore, a correction is needed to express these values in terms of total dietary amino acids, assuming protein in commonly-used feedstuffs is less than 100% digestible. For the present purposes, all of the maintenance requirement values were divided by an arbitrary apparent digestibility coefficient of 0.80 to arrive at the total amounts of dietary amino acids required for maintenance (Table 12.1).

Pregnancy

Amino acids are needed during pregnancy to replace those lost through obligatory sloughing or metabolism, to develop the pregnant uterus and its contents plus the mammary glands, and to add protein to the maternal body. Across sows, the most variable of these needs is that for maternal protein accretion, which ranges widely with differences in genotype, maturity and amount of maternal protein lost during a previous lactation.

Data were sought which would relate nitrogen accretion in the pregnant sow to intake of a limiting amino acid. Criteria for such a data set were:

1. Use of several levels of the limiting amino acid.
2. No combinations of intact proteins and crystalline amino acids, which might affect the efficiency of amino acid utilization by animals fed once daily a quantity of feed that would be consumed in a short time (Batterham and Murison, 1981).
3. Nitrogen balance measurements, using urinary catheters to minimize the serious overestimation of nitrogen retention often encountered in such experiments (Just, Fernendez and Jorgensen, 1982).

THREONINE

Only one data set meeting all of these criteria was found. King and Brown (1993) fed graded levels of a mixture of proteins to pregnant gilts and measured nitrogen balance (using urinary catheters) three times during pregnancy. By calculation, threonine was clearly the limiting amino acid. Daily threonine intake was regressed on daily nitrogen accretion at each measurement period. There was no evidence of curvature (reduced marginal efficiency) at higher threonine intake levels. Linear regressions produced coefficients of determination (R^2) of at least 0.99. Mean values for the intercept and slope across the three equations produced the following summary equation:

$$\text{Threonine intake (g)} = 1.45 + 0.53 \times \text{N accretion (g)}$$

The daily maintenance requirement of 1.45 g can be expressed in the case of these sows as 41mg/kg $W^{0.75}$, the value used in further calculations. From the slope (0.53) can be inferred a marginal (partial) efficiency of dietary threonine use for threonine accretion of 0.45. The maintenance requirement and slope can be used for estimation of the amount of threonine needed for a given animal and target mean daily nitrogen accretion during pregnancy, including nitrogen stored in the conceptus and mammary tissue.

OTHER AMINO ACIDS

It is assumed that the marginal efficiency of use of all amino acids for protein accretion is the same. This assumption may not be valid. However, there

Table 12.2   ESTIMATED AMINO ACID COMPOSITION OF BODY PROTEIN[a]

Amino Acid	g/16 g Nitrogen	AA:Thr[b]	AA:Lys[b]
Arginine	6.81	1.81	1.05
Histidine	2.91	0.77	0.45
Isoleucine	3.25	0.87	0.50
Leucine	7.12	1.89	1.09
Lysine	6.51	1.73	1.00
Methionine	1.75	0.47	0.27
Total sulphur amino acids	2.91	0.77	0.45
Phenylalanine	3.91	1.04	0.60
Total aromatic amino acids	6.73	1.79	1.03
Threonine	3.76	1.00	0.58
Tryptophan	0.68	0.18	0.10
Valine	4.50	1.20	0.69

[a] Data from Aumaitre and Duee (1974), Zhang *et al.* (1986) and Moughan and Smith (1987)
[b] Ratio of concentration of each aminio acid to threonine or to lysine.

are no data upon which to base an alternative hypothesis; some assumption must be made, and this one seems the simplest and safest. Given this assumption, the requirement for each other indispensable amino acid for protein accretion is taken to be in the same ratio to the threonine requirement for protein accretion as is the concentration ratio of that amino acid to threonine in body protein. The amino acid composition of body protein was estimated from three reports in the literature (Aumaitre and Duee, 1974; Zhang, Partridge and Mitchell, 1986; Moughan and Smith, 1987), and the results are presented in Table 12.2.

CALCULATION OF RECOMMENDATIONS

The amounts of indispensable amino acids needed during pregnancy are estimated by summing the amounts needed for replacing obligatory losses and for maternal protein accretion. Three items of information are needed for calculation of the recommended levels: (1) Weight at breeding; (2) anticipated total weight gain during pregnancy; and (3) target lean tissue accretion during pregnancy. The conversion from kg of lean tissue accretion during the entire pregnancy to grams of nitrogen retained per day was calculated assuming lean tissue is 23% protein (Williams, Close and Cole, 1985), protein is 16% nitrogen, and pregnancy lasts 115 days.

For example, consider the case of a gilt that weighs 110 kg at breeding and gains 60 kg total weight during pregnancy, for a mean body weight during pregnancy of 140 kg, corresponding to a metabolic body weight of 40.7 kg $W^{0.75}$. Replacement of obligatory loss of threonine is estimated to require 41 mg per kgW$^{0.75}$ or a total of 1.67g/day. The target lean tissue accretion from breeding to prefarrowing, including products of conception, is 40 kg, corresponding to daily retention of 12.8 g nitrogen. This 12.8 g nitrogen retained daily is multiplied by the slope (0.53 g threonine/g nitrogen retained) to show a need for 6.78 g threonine daily to support growth. The total amount of threonine to be supplied is 1.67 + 6.78 = 8.45 g/day.

The same sow requires 20 mg isoleucine/kg W$^{0.75}$ to replace obligatory loss, or a total of 0.81g/day. The ratio of isoleucine to threonine in body protein is 0.87 (Table 12.2), which is multiplied by 6.78 g threonine required daily to support growth, to show 5.87 g isoleucine needed daily for that purpose. The total amount of isoleucine to be supplied daily is 0.81 + 5.87 = 6.68 g. Requirements for other amino acids can be estimated in similar fashion, using the numbers in Tables 12.1 and 12.2. The lysine requirement is estimated to be 13.7 g/day.

Consider a very different animal, a multiparous sow bred at a body weight of 200 kg, and with a target lean tissue accretion, including products of conception, of 20 kg from breeding to prefarrowing. This rate of protein accretion is consistent with total gain from breeding to prefarrowing of about 35 kg, or from breeding to postfarrowing of about 15 kg. It corresponds to daily retention of 6.4 g nitrogen. Calculations similar to those above show this sow to need only 5.7 g threonine and 8.6 g lysine/day.

COMPARISON TO PREVIOUS ESTIMATES

These recommended amino acid intake levels during pregnancy appear slightly higher than some estimates of requirements. For example, the NRC (1988) values for both threonine (5.7 g/day) and lysine (8.2 g/day) and the ARC (1981) value for lysine (8.6 g/day) are consistent with only modest total lean tissue accretion during pregnancy, approximately 20 kg, as calculated in the present system. In the empirical experiments that form the basis of these previous requirement estimates, other factors including energy intake may have limited N retention and, thus, limited response to amino acid levels.

The balance of amino acids needed by pregnant sows can vary depending on the body weight and the rate of protein accretion (the relative demands for replacement of obligatory losses and for growth). The balance of amino acids calculated to be needed is in reasonable agreement with the balance specified by NRC (1988), except for two amino acids. Isoleucine is predicted to be needed in lesser amount and leucine in greater amount than proposed by NRC. Examination of the empirical data upon which the NRC levels are based is helpful. There is only one study of the response to each of these amino acids. The amount of isoleucine needed to maximize nitrogen retention was presented as a range (Kile and Speer, 1985). NRC chose the midpoint of the range, but the lower end of the range is in good agreement with the values calculated herein. Expression of the amount of leucine needed to maximize nitrogen retention (Easter and Baker, 1976) as g/day rather than as % of the diet, and correction for digestibility (the experiment used a crystalline amino acid diet) produces a value in reasonable agreement with the one estimated here.

MAMMARY DEVELOPMENT

Evidence from Australia (Head and Williams, 1992) suggests that inadequate amino acid intake during pregnancy may limit mammary development. The proposed method assumes that amino acid intake adequate for target protein accretion rates during pregnancy is also adequate for mammary development. Recent data from our laboratory raise questions about that assumption.

Kusina, Pettigrew, Sower, Crooker, White, Hathaway and Dial (1995) fed gestating sows maize-soybean meal diets with no crystalline amino acids to provide 4, 8 or 16 g lysine/day, then either a high or low amino acid intake during lactation. The intermediate gestation treatment supplied approximately the same amount of lysine as the requirement estimated by NRC (1988). Both milk yield and litter growth were markedly increased by increasing amino acid intake during either gestation or lactation, with no interaction. Among sows fed the higher amino acid levels during lactation, increasing lysine intake from 8 to 16 g/day during pregnancy increased average piglet weight gain during a 3-week nursing period from 4.0 to 4.8 kg.

Reduction of lactational performance by inadequate amino acid intake during pregnancy may have been mediated by any of three mechanisms: (1) Restricted development of the mammary glands; (2) limited maternal body protein reserves for mobilization to support milk production; and (3) impairment of piglet vigour. A part of the study not yet completed should help us understand whether mammary development was restricted by low intake of amino acids.

Regardless of the reason for the reduced lactational performance, it appears that effects on subsequent performance may need to be considered in estimating amino acid requirements for pregnant sows.

Lactation

Amino acids are needed by the lactating sow to replace obligatory losses and for synthesis of milk protein. They may be needed for glucoenogenesis in some situations, but that is not considered in the present calculations. The response of lactating sows to dietary amino acid intake is complicated by the mobilization of the sow's body protein stores to support milk protein synthesis. This mobilization is driven by strong lactational homeorhesis, orchestrated action of many organs to achieve the goal of milk production (Bauman and Currie, 1980). This homeorhetic drive may be stronger in some sows than in others, suggesting variation in the response of sows to amount of dietary lysine. Furthermore, massive differences in milk yield exist among sows.

LYSINE

Estimates of the lysine needs of lactating sows to maximize litter growth rate range from less than 20 g/day (Boomgaardt, Baker, Jensen and Harmon, 1972) to more than 50 g/day (Johnston, Pettigrew and Rust, 1993). Examination of seven published estimates of requirements reveals a close association between amount of lysine apparently needed to maximize lactational performance and the level of that performance. Regressing these seven estimates (in some cases my estimates differed from those suggested in the original publication) against the daily litter growth rate on that treatment gives the following relationship:

$$\text{Lysine, g/day} = -6.71 + 0.026 \times \text{litter growth, g/day}$$

The R^2 is 0.77, the standard error of the slope is 0.006, and that of the estimate is 6.43.

The negative intercept (−6.71) reflects the homeorhetic mobilization of body protein to support milk protein synthesis. The slope (0.026) indicates that an increment of 1 kg of litter growth per day requires an additional 26 g of dietary lysine per day. No deviation from linearity is obvious. This slope can be calculated to reflect a marginal efficiency of conversion of dietary lysine to lysine in the body protein of suckling piglets of 0.40.

The relationship above is subject to error because of inappropriate selection of the lysine requirement in individual experiments. Therefore, a separate regression was calculated that included all 25 of the treatments in which lysine was not provided in excess of the requirement (i.e. some other factor limited litter growth) in the same experiments. The slope was identical to that in the equation given above, and the intercept was only slightly more negative (−8.38). The close agreement between these two estimates of the effect of potential litter growth on lysine needs lends confidence.

The response of lactating first-parity sows to intake of amino acids (expressed as lysine) and metabolizable energy (ME) has been reported (Tokach, Pettigrew, Crooker, Dial and Sower, 1992). A strong interaction between lysine and ME, in their effect on lactational performance was found, consistent with the relationships described above. Sows given a liberal supply of ME responded to much higher levels of lysine intake than did sows with a severely restricted ME intake. A regression similar to those above was calculated for the seven treatments in which lysine supply was not clearly in excess of needs. It produced a slope of 0.028, encouragingly similar to that reported above. However, the intercept was much more strongly negative (−25.30), perhaps because of stronger lactational home-orhesis. Whatever its cause, the large difference in intercept suggests that the requirements presented in this review, calculated from the equation shown above, may seriously overestimate the lysine needs of some sows for a stated rate of litter growth. These sows would presumably produce more litter growth than otherwise expected by taking more protein from the body stores than do the 'typical' sows upon which the first relationship is based.

Among the genetic factors that influence response of lactating sows to amino acid intake, variation in the amount of milk produced is presumably the most important. However, the strength of lactational homeorhesis, i.e. the drive to mobilize body tissues to support milk production, may be second in importance. Unfortunately, far too little information is available to allow that variation in homeorhesis to be considered when setting nutrient levels.

The logic, then, upon which the recommended lysine supply for lactating sows in this review is based is that the amount needed varies with the level of litter growth possible if lysine supply is ample. This level of litter growth can be restricted by genotype, environment, or intake of other nutrients.

OTHER AMINO ACIDS

Three factors are considered in estimating the amino acid needs during lactation. First is the requirement for replacement of obligatory losses, calculated as it is for pregnancy. Second is the amount needed to support milk production. The need for lysine for milk production is a function of the litter growth rate and the slope of the equation given above (0.026g lysine/g litter growth). Needs for other amino acids for milk production are assumed to be the same relative to lysine as their concentration ratios in milk protein, an assumption that we now know to be invalid as described later. Third is the amount provided by mobilization of body protein, taken as a constant for purposes of these calculations. For lysine this quantity is the difference between the amount needed for maintenance and the intercept in the equation above (−6.71 g/day). If the mean body weight of sows in the experiments that contributed data to the regression analysis is assumed to be 155 kg, and the maintenance requirement for lysine given in Table 12.1 is used, the daily maintenance requirement is estimated as 2.14 g/day. This reveals a total lysine from protein mobilization of 2.14 − (−6.71) = 8.85 g/day. The supply of other amino acids from mobilized protein is assumed to be in the same ratio to lysine as in body protein (Table 12.2).

One of the key features of these estimates of amino acid needs of lactating sows is that they do not make the unrealistic assumption that lactating sows are at zero nitrogen balance. It is assumed that lactating sows are in negative nitrogen balance. However, the available information will not support calculation of varying rates of protein mobilization, forcing the assumption, for calculation purposes, that the amount of amino acid derived from body protein is constant. Increasing amino acid intake above the present recommendations would probably reduce protein loss from the sow's body.

An estimate of amino acid composition of sows' milk protein, derived from the data of Elliott, Van der Noot, Gilbreath and Fisher (1971), Duee and Jung (1973) and King (unpublished), is shown in Table 12.3.

Table 12.3 ESTIMATED AMINO ACID COMPOSITION OF SOWS' MILK PROTEIN[a]

Amino Acid	g/16g Nitrogen	AA:Lys[b]
Arginine	4.93	0.66
Histidine	3.02	0.40
Isoleucine	4.11	0.55
Leucine	8.6	1.15
Lysine	7.5	1.00
Methionine	1.94	0.26
Total sulphur amino acids	3.40	0.45
Phenylalanine	4.12	0.55
Total aromatic amino acids	8.42	1.12
Threonine	4.38	0.58
Trytophan	1.32	0.18
Valine	5.50	0.73

[a] Data from Elliott *et al.* (1971), Duee and Jung (1973) and R.H. King (unpublished)
[b] Ratio of concentration of each amino acid to lysine.

CALCULATION OF RECOMMENDATIONS

An example sow farrows at body weight of 160 kg, and loses 20 kg during lactation. Assuming a constant rate of loss, the average body weight during lactation is 150 kg, corresponding to metabolic body weight of 42.9 kg $W^{0.75}$. Daily obligatory losses of lysine are estimated to be 2.09 g/day. Litter growth rate is 2 kg/day. The slope in the equation above shows a requirement of 26 g lysine/kg litter growth, for a total requirement of 52.00 g lysine/day to support litter growth. The amount obtained from mobilization of body stores is assumed to be a constant 8.85 g/day. The total dietary need for lysine is therefore 2.09 + 52.00 − 8.85 = 45.24 g/day.

The same sow needs 0.86 g isoleucine/day to replace obligatory losses, calculated from the maintenance requirement given in Table 12.1. The ratio of isoleucine to lysine in milk protein is 0.55, which is multiplied by 52.00 g lysine to show 28.51 g isoleucine required daily to support milk synthesis. The ratio of isoleucine to lysine in body protein mobilized is 0.50, which is multiplied by 8.85 to show 4.43 g of isoleucine provided daily from the sow's body. The total amount of isoleucine needed daily is then 0.86 + 28.51 − 4.43 = 24.94 g.

If the litter growth rate were 2.5 kg/day, the requirements would be 58.2 g lysine and 32.1 g isoleucine/day. If the litter gained only 1.5 kg/day, the daily requirements would be only 32.2 g lysine and 17.8 g isoleucine.

COMPARISON TO PREVIOUS ESTIMATES

These recommendations are higher than those of NRC (1988), at least for very productive sows. The NRC estimate of the lysine requirement is 31.8 g/day, corresponding in the present calculations to a modest litter growth rate of 1460 g/day. The present recommendations are, however, consistent with the results of the several experiments upon which the original regression equation was based.

Comparison of the balance of amino acids derived by the present method to that derived from empirical studies (NRC, 1988) shows good agreement except for leucine and valine. The present calculations indicate considerably more leucine and less valine are needed than does NRC. If the NRC value for leucine is correct, either the leucine content of milk protein is overestimated or leucine is used at a dramatically higher efficiency for milk protein synthesis than are the other amino acids. Inspection of the only empirical study upon which the NRC estimate is based (Rousselow, Speer and Haught, *et al.*, 1979) shows a very low level of lactational performance

(<1200 g litter gain/day). Perhaps sows producing more milk would have responded to higher levels of leucine.

VALINE

It is now clear that the valine requirement to maximize milk yield is higher than estimated by the method proposed herein. This method requires the assumption that all amino acids are used with equal efficiency by the mammary gland in the synthesis of milk protein. There is strong evidence at both the tissue and whole-animal levels that valine is used less efficiently for milk synthesis than are most other amino acids. A considerable amount of valine is oxidized to CO_2 by goat mammary glands (Roets, Massmart-Leen, Verbeke and Peeters, 1979). Arterio-venous difference experiments in cows (Wohlt, Clark, Derrig and Davis, 1977) and pigs (Linzell, 1969; Trottier, Shipley and Easter, 1994) suggest that the ratio of valine taken up by the mammary gland to the amount secreted in milk is greater than corresponding ratios for other amino acids. The first empirical estimation of the valine requirement of lactating sows (Rousselow and Speer, 1980) found a higher requirement than had previously been estimated on the basis of amino acid compositon of milk protein. A more recent experiment (Tokach, Goodband, Nelssen and Kats, 1993) showed that with diets containing 9.0 g lysine/kg, raising the valine level above 7.5 g/kg increased litter weaning weights, and that the effect was greater in larger litters. The valine/lysine ratio in the control diet was 0.83, higher than the ratio of 0.73 estimated by the proposed method. Most recently, we at the University of Minnesota have collaborated with researchers at Kansas State University in conducting a feeding trial to estimate the valine requirement of lactating sows (Richert, Goodband, Tokach, Nelssen, Johnston, Walker, Pettigrew and Blum, 1994). All diets contained 9.0 g lysine/kg, and all other amino acids except valine were present in adequate amount. Selected results are shown in Table 12.4. Litter growth increased with increasing valine to at least 10.5 g/kg, and perhaps to 11.5 g/kg. These levels are 1.17 and 1.28, respectively, times the lysine level.

Overall, it is clear that the valine requirement for lactating sows is greater than estimated by this method, because the method has no mechanism to adjust estimates for the oxidation of valine in the mammary gland. The valine requirement of high-producing sows is greater than the lysine requirement, and probably about 20% higher. This may have important effects on lactation diet formulation.

Table 12.4   RESPONSE OF LACTATING SOWS TO DIETARY VALINE
CONCENTRATION

| | *Dietary Valine, g/kg* | | | | | |
Item	7.5	8.5	9.5	10.5	11.5	cv
Feed intake, kg/day	6.42	6.00	6.15	6.34	6.27	14.30
Pigs weaned/litter	10.20	10.20	10.10	10.20	10.20	3.70
Litter weight, kg						
Day 21[b]	62.44	62.55	63.97	64.97	65.53	11.90
Weaning[bc]	76.19	76.13	77.77	78.80	79.92	11.00
Litter weight gain, kg						
Day 0 to 7[d]	12.23	12.70	12.70	12.89	13.24	18.00
Day 0 to 21[b]	46.86	47.07	48.30	49.51	49.60	13.90
Day 0 to weaning[bc]	60.60	60.66	62.09	63.34	63.99	12.60

[a] From Richert *et al.* (1994). All diets contained 9.0g lysine/kg
[b] Linear effect of valine, $P < 0.06$.
[c] Mean weaning age 26 days
[d] Linear effect of valine, $P < 0.06$.

SUBSEQUENT REPRODUCTION

This method of estimating amino acid requirements of lactating sows relies on the assumption that amino acid intake adequate to maximize litter growth would also maximize subsequent reproduction. Recent data from Australia (Campbell and King, unpublished) suggest that the subsequent litter size may be dramatically increased by increasing amino acid intake of lactating primiparous sows above the level required for maximum litter growth. This is an extremely important finding. Assuming it is verified by further experiments, it will be important to estimate the quantitative requirements to maximize subsequent litter size, and to determine whether this phenomenon is limited to young sows.

This result shows again that effects of amino acid intake on subsequent performance must be considered.

Digestible amino acids

The description above is of estimation of total dietary amino acid requirements, but the proposed method can be used also to estimate apparent ileal digestible amino acid requirements.

First, the maintenance requirements, except for threonine, are set at the original estimates of Fuller *et al.* (1989). They may be easily obtained by multiplying the values in Table 12.1 by 0.80. The maintenance requirement for threonine is 38 mg/kg $W^{0.75}$, obtained from the intercept of a new regression equation.

Regression equations presented earlier for total dietary amino acids were recalculated, using estimates of apparent ileal digestible threonine or lysine (as appropriate) in the original experimental diets. For pregnant sows, the slope is 0.40 g digestible threonine/g nitrogen accretion. For lactating sows, the slope is 23 g digestible lysine/kg litter gain. The constant amount of lysine derived from the body is estimated in this case to be 9.07 g/day. The amino acid ratios in body protein and in milk are the same as used in estimation of total dietary amino acid requirements.

Final comments

It is important in deriving amino acid levels for a specific situation that actual production levels be known and used. It is not helpful to assume a litter growth rate of 2 kg/day if the energy intake of lactating sows is inadequate to support that rate of litter growth. It is also important to ensure that potential litter growth rates are measured in the presence of adequate amino acid intake.

The recommended amino acid levels are all given in g/day. They can be easily converted to % of diet for formulation purposes by dividing them into the actual feed intake in a specific situation. Reliable information on the amount of feed consumed by the sows is essential. The amount consumed is not necessarily the amount targeted for pregnant sows or the amount desired for lactating sows. The production record systems that are so important in overall management of pork production should provide the information needed for use of these recommendations. The professionalism and efficiency required of the pork industry cannot be achieved without use of adequate record systems.

The recommendations presented in this document should be used only until better ones are available. Better recommendations will come from more information on the biological responses of sows to amino acid levels, and from more thorough integration of available information.

One of the frustrations encountered in developing recommended nutrient levels comes from the fact that our qualitative knowledge of biological phenomena at organ and cellular levels goes far beyond the data available to allow those phenomena to be incorporated quantitatively into recommendations. That is the case presently with regard to amino acid nutrition of sows, and so the recommendations in this document are considerably less sophisticated than are our concepts of amino acid nutrition of sows. Present data do not allow thorough consideration of different efficiencies of use of different amino acids, or of the same amino acid in different conditions; of the effects of body reserves on lactational performance; or of the effects of amino acid nutritional status on reproductive performance or on mammary development. More sophisticated modelling approaches will be required to consider these important factors.

Summary

A method for estimating amino acid requirements of sows is presented. It offers the important advantage of being responsive to variations in level of production and in body weight, but is firmly anchored to empirical data. It is imperfect, but suggested to be superior to simpler systems. It has stimulated research into areas where information is lacking. Results will provide the information needed to improve it. Valine requirements of lactating sows are higher than predicted by this method, and effects of amino acid nutrition during one stage of the reproductive cycle on subsequent performance need thorough evaluation.

References

Agricultural Research Council (1981) *The Nutrient Requirements of Farm Livestock, No. 3: Pigs.* London: Agricultural Research Council

Aumaitre, A., and Duee, P.H. (1974) Composition en acides amines des proteines corporelles du porcelet entre la naissance et l'age de huit semaines, *Annales de Zootechnie,* **23**, 231–236

Batterham, E.S., and Murison, R.D. (1981) Utilization of free lysine by growing pigs, *British Journal of Nutrition,* **46**, 87–92

Bauman, D.E., and Currie, W.B. (1980) Partitioning of nutrients during pregnancy and lactation: A review of the mechanisms involving homeostasis and homeorhesis, *Journal of Dairy Science,* **63**, 1514–1529

Boomgaardt, J., Baker, D.H., Jensen, A.H., and Harmon, B.G. (1972) Effect of dietary lysine levels on 21-day lactation performance of first-litter sows, *Journal of Animal Science,* **34**, 408–410

Dourmad, J.Y., Etienne, M., and Noblet, J. (1991) Contribution a l'etude des besoins en acides amines de la truie en lactation, *Journees Recherche Porcine en France,* **23**, 61–68

Duee, P.H., and Jung, J. (1973) Composition en acides amines du lait de truie, *Annales de Zootechnie*, **22**, 243–247

Easter, R.A., and Baker, D.H. (1976) Nitrogen metabolism and reproductive response of gravid swine fed an arginine-free diet during the last 84 days of gestation, *Journal of Nutrition*, **106**, 636–641

Elliot, R.F., Van der Noot, G.W., Gilbreath, R.L., and Fisher, H. (1971) Effect of dietary protein levels on composition changes in sow colostrum and milk, *Journal of Animal Science*, **32**, 1128–1137

Fuller, M.F., McWilliam, R., Wang, T.C., and Giles, L.R. (1989) The optimum dietary amino acid pattern for growing pigs, 2. Requirements for maintenance and for tissue protein accretion, *British Journal of Nutrition*, **62**, 255–267

Head, R., and Williams, I.H. (1992) Mammogenesis is influenced by pregnancy nutrition, In: *Manipulating Pig Production III*. Edited by E.S. Batterham, Attwood, Victoria, Australia: Australasian Pig Science Association

Johnston, L.J., Pettigrew, J.E., and Rust, J.W. (1993) Response of maternal-line sows to dietary protein concentration during lactation, *Journal of Animal Science*, **71**, 2151–2156

Just, A., Fernendez, J.A., and Jorgensen, H. (1982) Nitrogen balance studies and nitrogen retention. *Physiologie Digestive Chez le Porc*. Les colloques de l'INRA, no. 12

Kile, D.L., and Speer, V.C. (1985) Isoleucine requirement for reproduction in swine. *Journal of Animal Science*, **61** (Suppl. 1), 103 (Abstr.)

King, R.H., and Brown, W.G. (1993) Interrelationships between dietary protein level, energy intake, and nitrogen retention in pregnant gilts. *Journal of Animal Science*, **71**, 2450–2456

Kusina, J., Pettigrew, J.E., Sower, A.F., Crooker, B.A., White, M.E., Hathaway, M.R., and Dial, G.D. (1995) The effect of protein (lysine) intake during gestation and lactation on lactational performance of the primiparous sow. *Journal of Animal Science*. **73** (Suppl. 1), Submitted (Abstr.)

Linzell, J.L., Mepham, T.B., Annison, E.F., and West, C.E. (1969) Mammary metabolism in lactating sows: arteriovenous differences of milk precursors and the mammary metabolism of [^{14}C] glucose and [^{14}C] acetate, *British Journal of Nutrition*, **23**, 319–332

Moughan, P.J., and Smith, W.C. (1987) Whole-body amino acid composition of the growing pig. *New Zealand Journal of Agricultural Research*, **30**, 301–303

National Research Council. (1988) *Nutrient Requirements of Swine*. Ninth Revised Edition, Washington, D.C.: National Academy Press

Pettigrew, J.E. (1993) Amino acid nutrition of gestating and lactating sows. *Biokyowa Technical Review – 5.* Chesterfield, MO: Nutri-Quest Inc.

Richert, B.T., Goodband, R.D., Tokach, M.D., Nelssen, J.L., Johnston, L.J., Walker, R.D., Pettigrew, J.E., and Blum, S.A. (1994) Determining the valine requirement of high producing sows. *Journal of Animal Science,* **72** (Suppl. 1), 389 (Abstr.)

Roets, E., Massart-Leen, A.M., Verbeke, R., and Peeters, G. (1979) Metabolism of [U-^{14}C; 2,3-^{3}H]-L-valine by the isolated perfused goat udder. *Journal of Dairy Research,* **46**, 47–57

Rousselow, D.L., Speer, V.C., and Haught, D.G. (1979) Leucine requirement of the lactating sow, *Journal of Animal Science,* **49**, 498–506

Rousselow, D.L., and Speer, V.C. (1980) Valine requirement of the lactating sow, *Journal of Animal Science,* **50**, 472–478

Tokach, M.D., Pettigrew, J.E., Crooker, B.A., Dial, G.D., and Sower, A.F. (1992) Quantitative influence of lysine and energy intake on yield of milk components in the primiparous sow. *Journal of Animal Science,* **70**, 1864–1872

Tokach, M.D., Goodband, R.D., Nelssen, J.L., and Kats, L.J. (1993) Valine – a deficient amino acid in high lysine diets for the lactating sow. *Journal of Animal Science,* **71** (Suppl. 1), 68 (Abstr.)

Trottier, N.L., Shipley, C.F., and Easter, R.A. (1994) Arteriovenous differences for amino acids, urea nitrogen, ammonia and glucose across the mammary gland of the lactating sow, *Journal of Animal Science,* **72** (Suppl. 1), 332 (Abstr.)

Williams, I.H., Close, W.H., and Cole, D.J.A. (1985) Strategies for sow nutrition: Predicting the response of pregnant animals to protein and energy intake, In *Recent Advances in Animal Nutrition – 1985,* pp. 133–147, Edited by W. Haresign and D.J.A. Cole, London: Butterworths

Wohlt, J.E., Clark, J.H., Derrig, R.G., and Davis, C.L. (1977) Valine, leucine and isoleucine metabolism by lactating bovine mammary tissue, *Journal of Dairy Science* **60**, 1875–1882

Zhang, Y., Partridge, I.G., and Mitchell, K.G. (1986) The effect of dietary energy level and protein:energy ratio on nitrogen and energy balance, performance and carcass composition of pigs weaned at 3 weeks of age. *Animal Production,* **42**, 389–395

LIST OF PARTICIPANTS

The twenty-ninth Feed Manufacturers Conference was organized by the following committee:

Miss P.J. Brooking (W J Oldacre Ltd)
Dr W.H. Close (Close Consultancy)
Dr J. Harland (Trident Feeds)
Dr S. Jagger (Dalgety Agriculture)
Dr D. Kitchen (Amalgamated Farmers Ltd)
Mr D.R. McLean (W L Duffield & Sons Ltd)
Mr R.T. Pass (United Distillers)
Mr J.R. Pickford
Dr A. Reeve (ICI Nutrition)
Mr E. Ross (BOCM Pauls)
Mr D.H. Thompson
Mr J.Twigge (Trouw Nutrition)
Dr K.N. Boorman
Prof P.J. Buttery
Dr D.J.A. Cole (Chairman)
Dr P.C. Garnsworthy (Secretary)   }   University of Nottingham
Dr W. Haresign
Prof G.E. Lamming
Dr J. Wiseman

The conference was held at the University of Nottingham, Sutton Bonington Campus, 4th-6th January 1995 and the committee would like to thank the authors for their valuable contributions. The following persons registered for the meeting:

Abbott, Mrs T.	Trouw Nutrition (UK) Ltd, Wincham, Northwich, Cheshire, CW9 6DF
Adams, Dr C.	Kemin Europa NV, Industrie Zone Wolfstee, 2410 Herentals, Belgium
Albers, Dr N.	BASF A6, Mevila D205, D67056 Ludwigshafen, Germany
Alderman, Mr G.	Nutrtion Consultant, Hunters Moon, Pearmans Glad, Shinfield Road, Reading RG2 9BE
Allan, Dr J.D.	Frank Wright Ltd, Ashbourne, Derbyshire, DE6 1HA
Allder, Mr M.	Eurotech Nutrition Ltd, B Martins Land, Witcham, Ely, Cambs CB6 2LB
Allen, Miss D.	Genus Management, 1 Colmer Road, Yeovil, Somerset
Anderson, Mr K.	WL Duffield & Sons Ltd, Saxlingham Thorpe Mills, Norwich, Norfolk, NR15 1TY
Angold, Mr M.	Roche Products Ltd, Heanor Gate, Heanor, Derby
Aronen, Dr I.	Rehuraisio OY, PL 101, FIN-21201, Raisio, Finland
Ashington, Mr B.	Trouw Nutrition (UK) Ltd, Wincham, Northwich, Cheshire, CW9 6DF
Aspland, Mr P.	Aspland & James Ltd, Bridge Street, Chatteris, Cambs, PE16 6RN
Atherton, Dr D.	Thomson & Joseph Ltd, 119 Plumstead Road, Norwich
Auran, Dr T.	Felleskjopet Forutikling, Postbox 3771, 7002 Trondheim, Norway
Barrie, Mr M.J.	Elanco Animal Health, Dextra Court, Chapel Hill, Basingstoke, Hampshire RB21 2SY
Bartram, Dr C.	Dalgety Agriculture Ltd, 180 Aztec West, Almondsbury, Bristol, BS12 4TH
Bates, Ms A.	Vitrition, Ryhall Road, Stamford, Lincs, PE9 1TZ
Beard, Mr M.	University of Nottingham, Sutton Bonington Campus, Loughborough, Leics, LE12 5RD
Bedford, Dr M.R.	Finnfeeds International Ltd, Ailesbury Court, High Street, Marlborough, Wilts

Beer, Mr J.H.	W & J Pye Ltd, Fleet Square, Lancaster, LA1 1HA
Beesty, Mr C.	Lloyds Animal Feeds Ltd, Morton, Oswestry, Shropshire
Bell, Miss J.F.	W & J Pye Ltd, Fleet Square, Lancaster, LA1 1HA
Bercovici, Mr D.	Eurolysine, 16 Rue Ballu, 75009 Paris, France
Berry, Mr M.H.	Berry Feed Ingredients Ltd, Chelmer Mills, New Street, Chelmsford, Essex CM1 1PN
Best, Mr P.	Feed International, 18 Chapel Street, Petersfield, Hants
Blake, Dr J.	Highfield, Little London, Andover, Hants, SP11 6JE
Bole, Mr J.	David Patton Ltd, Milltown Mills, Monaghan, Ireland
Boorman, Dr K.N.	University of Nottingham, Sutton Bonington Campus, Sutton Bonington, Loughborough, Leics LE12 5RD
Booth, Miss A.	Yorkshire County Feeds, Darlington Road, Northallerton Road, Northallerton, DL6 2NU
Borgida, Mr L.P.	COFNA, 25 rue du Rempart, 37018 Tours, Cedex, France
Bouchard, Mr K.	Midland Shires Farms Ltd, County Mills, Worcester, NR1 3NU
Bourdillon, Mrs A.	Sanders-Aliments, 17 quai de l'Industrie, 91200-Athis-Mons, France
Bourne, Mr S.	Alltech (UK) Ltd, 16117 Abenbury Way, Wrexham Ind. Estate, Wrexham, Clwyd LL13 9UZ
Brenninkmeijer, Dr C.	Hendrix' Voeders BV, Postbus 1, 5830 MA Boxmeer, The Netherlands
Brooking, Miss P.	International Additives, Old Gorsey Lane, Wallasey, Merseyside
Brophy, Mr A.	Alltech Ireland Ltd, 28 Cookstown Industrial Estate, Tallaght, Dublin 24
Brown, Mr G.J.P.	Colborn Dawes Nutrition, Heanorgate, Heanor, Derbyshire
Brown, Mr J.M.	Britphos Ltd, Rawson House, Yeadon, Leeds

Burt, Dr A.W.A.	Burt Research Ltd, Kimbolton, Huntingdon, Cambs PE18 0HU
Butter, Miss N.	University of Nottingham, Sutton Bonington Campus, Loughborough, Leics LE12 5RD
Buttery, Prof P.J.	University of Nottingham, Sutton Bonington Campus, Loughborough, Leics LE12 5RD
Buysing Damste, Dr B.	BP Chemicals Ltd, PO Box 60, 4000 AB Tiel, The Netherlands
Campani, Dr I.	F.LLI Martini e C p.a., 4020-Longiano-Fo, Italy
Carmichael, Mr D.C.	Elanco Animal Health, Dextra Court, Chapel Hill, Basingstoke, Hampshire RG21 2SY
Carter, Mr A.	Hoechst Animal Health, Walton Manor, Walton, Milton Keynes, MK7 7AJ
Carter, Mr T.	Kemin (UK) Ltd, Becor House, Green Lane, Lincoln LN6 7DL
Chamberlain, Dr D.	Hannah Research Institute, Ayr, Scotland, KA6 5HL
Charles, Dr D.R.	ADAS Nottingham, Chalfont Drive, Nottingham, NG8 3SN
Charlton, Mr P.	Alltech (UK) Ltd, 16117 Abenbury Way, Wrexham Ind Estate, Wrexham, Clwyd LL13 9UZ
Chase, Dr L.E.	Cornell University, Dept Animal Sciences, 149 Morrison Hall, Ithaca, NY 14853-4801 USA
Chew, Dr B.P.	Washington State Univeristy, Dept Animal Science, 261 Clark Hall, Pullman, WA 99164-6320 USA
Chignola, Dr P.	V Valpantena 18G, 37034-Quinto, Verona, Italy
Choung, Dr J.J.	Hannah Research Institute, Ayr, KA6 5HL
Clay, Mr J.	Alltech (UK) Ltd, 16117 Abenbury Way, Wrexham Ind Estate, Wrexham, Clwyd LL13 9UZ
Close, Dr W.H.	Close Consultancy, 129 Berkham Road, Wokingham, Berks, RG11 2RG
Cole, Dr D.J.A.	University of Nottingham, Sutton Bonington Campus, Loughborough, Leics LE12 5RD

Cole, Mr J.	International Additives, Old Gorsey Lane, Wallasey, Merseyside
Colenso, Mr J.	Trouw Nutrition (UK) Ltd, Wincham, Northwich, Cheshire, CW9 6DF
Connolly, Mr J.G.	Red Mills Ltd, Goresbridge, Co Kilkenny, Ireland
Cooke, Dr B.	Dalgety Agriculture Ltd, 180 Aztec West, Almondsbury, Bristol, BS12 4TH
Cooper, Mr A.	Seale-Hayne, Newton Abbot, Devon, TQ12 6NQ
Cottrill, Dr B.	ADAS, Woodthorne, Wergs Road, Wolverhampton WV6 8TQ
Cowan, Dr D.	Novo Nordisk SA, 282 Chartridge Lane, Chesham, Bucks
Cox, Mr N.	SCA Nutrition Ltd, Maple Mill, Dalton, Thirsk
Creasey, Mrs A.	BASF Plc, Box 4, Earl Road, Cheadle Hulme, SK8 6QG
Cullin, Mr A.W.R.	Forum Holdings, Forum House, 41-52 Brigton Road, Redhill
Dallas, Mr J.L.	Crediton Milling CoLtd, Fordton Mill, Crediton, Devon, EX17 3DH
Dawson, Mr W.	Britphos Ltd, Rawdon House, Green Lane, Yeadon, Leeds LS19 7BY
De Lange, Mr L.L.M.	Cavo Latuco, PO Box 8210, 3503 RE Utrecht, Holland
Deaville, Mr S.P.	Rumenco, Stretton House, Burton on Trent, Staffs, DE13 0DW
Diepenbroek, Mr L.D.	Mole Valley Farmers, Station Road, South Molton, Devon
Dixon, Mr D.H.	Brown & Gillmer Ltd, PO Box 3154, The Lodge, 199 Strand Road, Merrion, Dublin 4
Doran, Mr D.	Trouw Nutrition (UK) Ltd, Wincham, Northwich, Cheshire CW9 6DF
Drakley, Miss C.	University of Nottingham, Sutton Bonington Campus, Loughborough, Leics LE12 5RD
Eclache, Dr D.	Laboratories Pancosma SA, 6 Voie-des-Traz/cp 143, CH-1218 Grand-Saconnex, Geneva, Switzerland

Edwards, Miss S.	Trouw Nutrtion (UK) Ltd, Wincham, Northwich, Cheshire CW9 6DF
Ewing, Mrs A.	Dalgety Agriculture Ltd, 180 Aztec West, Almondsbury, Bristol BS12 4TH
Ewing, Dr W.	Cargill plc, Camp Road, Swinderby, Lincoln LN6 9TN
Fawcett, Mr T.	AF Plc, Kinross, New Hall Lane, Preston, Lancs
Filmer, Mr D.	David Filmer Ltd, Wascelyn, Brent Knoll, Somerset, TA9 4DT
Fitt, Dr T.	Roche Products Ltd, Heanor Gate, Heanor, Derbys
Fitzsimmons, Mr J.	Volac Ltd, Orwell, Royston, Herts, SG8 5QX
Fletcher, Mr C.J.	Aynsome Laboratories, Eccleston Grange, Presdcot Road, St Helens, Merseyside
Ford, Mr M.	Hydro Chafer Ltd, York Road, Elvington, York
Fordyce, Mr J.	West Midlands Farmers Ltd, Bradford Road, Melksham, Wilts SN12 8LQ
Forster, Mr P.	Optivite Ltd, Main Street, Laneham, Retford, Nottingham
Foulds, Mr S.	Park Tonks Ltd, Abingdon House, 48 North Road, Gt Abingdon, Cambs CB1 6AS
Frumholtz, Dr P.P.	1 Place Charles de Gaulle, BP 301, 78054 St Quentin en Yvelimes, France
Fullarton, Mr P.J.	Forum Chemicals Ltd, 2 Hookstone Park, Harrogate, Yorks HG2 8QT
Gaisford, Mr M.	Farmers Weekly, Quadrant House, Sutton, Surrey SM2 5AS
Garnsworthy, Dr P.C.	University of Nottingham, Sutton Bonington Campus, Loughborough, Leics LE12 5RD
Geddes, Mr N.	Nutec Ltd, Eastern Avenue, Lichfield, Staffs
Gibson, Mr J.E.	Parnutt Foods Ltd, Hadley Road, Woodbridge Industrial Estate, Sleaford, Lincs NG34 7EG
Gill, Dr P.	Meat and Livestock Commission, PO Box 44, Winter Hill House, Snowdon Drive, Milton Keyns MK6 1AX

Gillespie, Miss F.	United Molasses, Stretton House, Derby Road, Stretton, Burton on Trent, Staffs, DE13 9DW
Givens, Dr D.I.	ADAS Feed Evaluation Unit, Alcester Road, Stratford-on-Avon
Goff, Mr S.	BMS Computer Solutions Ltd, Sproughton House, Sproughton, Ipswich, Suffolk IP8 3AW
Gooderham, Mr B.	W&J Pye Ltd, Fleet Square, Lancaster, LA1 1HA
Gore, Mr A.	University of Nottingham, Sutton Bonington Campus, Loughborough, Leics LE12 5RD
Gould, Mrs M.	Volac Ltd, Orwell, Royston, Herts SG8 5QX
Grace, Mr J.R.	Elanco Animal Health, Dextra Court, Chapel Hill, Basingstoke, Hampshire RG21 2SY
Gray, Mr W.	Kemira Kemi (UK) Ltd, Orm House, 2 Hookstone Park, Harrogate, HG2 8QT
Griffiths, Mr W.D.E.	MSF Ltd, Defford Mill, Earls Croome, Worcester WR8
Hall, Mr A.C.	R Keenan (UK) Ltd, Harters Hill Lane, Coxley, Wells, Somerset BA5 1RD
Haresign, Dr W.	University of Nottingham, Sutton Bonington Campus, Loughborough, Leics LE12 5RD
Harker, Dr A.J.	Finnfeeds International Ltd, Market House, Ailesbury Court, High Street, Malborough Wilts SN8 1AA
Harris, Mr C.	18 Priest Hill, Caversham, Reading, Berks RG4 7RZ
Harrison, Mrs J.	Sciantec Analytical Services, Main Site, Dalton, Thirks, N Yorks Y01 3JA
Harrison, Mr W.M.	Fosse Ltd, Whetstone Magna, Lutterworth Road, Whetstone, Leics LE8 6NB
Haythornthwaite, Mr A.	Nu Wave, 45 Church Road, Warton, Preston, PR4 1BD
Hazledine, Mr M.	Dalgety Agriculture Ltd, 180 Aztec West, Almondsbury, Bristol BS12 4TH
Heavey, Mr S.	Stewarts Feeds, Boyle, Co Roscommon, Ireland
Hellberg, Mr S.	LFU, Box 30192, 104 25 Stockholm, Sweden

Hemke, Mr G.	PO Box 200, 5460 BC Veghel, The Netherlands
Heppinstall, Miss B.	Trouw Nutrition (UK) Ltd, Wincham, Northwich, Cheshire, CW 9 6DF
Hewson, Mr R.	UFAC (UK) Ltd, Waterwitch House, Exeter Road, Newmarket, CB8 8LR
Higginbotham, Dr J.D.	United Molasses, Stretton House, Derby Road, Stretton, Burton on Trent, Staffs DE13 0DW
Hocke, Miss N.	Lucta SA, PO Box 1112, 08080 Barcelona, Spain
Hockey, Dr R.	Smith Beacham Ltd, Hunters Chase, Walton Oaks, Dorking Road, Tadworth, Surrey KT20 7NT
Hogg, Mr A.A.	Elanco Animal Health, Dextra Court, Chapel Hill, Basingstoke, Hampshire RG21 2SY
Holmes, Mr G.F.	SE Johnson Ltd, Old Road Mills, Darley Dale, Matlock, Derbys DE4 2ES
Howie, Mr A.D.	NutritionTrading (Int) Ltd, Morton Bagot, Studley, Warks, B80 7ED
Hughes, Mr D.P.	NWF Agriculture Ltd, Wardle, Nantwich, Cheshire CW5 6AQ
Hunt, Miss L.	University of Nottingham, Sutton Bonington Campus, Loughborough, Leics LE12 5RD
Hurley, Miss J.	Dalgety Agriculture Ltd, 180 Aztec West, Almondsbury, Bristol, BS12 4TH
Ince, Mr R.G.	BMS Computer Solutions Ltd, Sproughton House, Sproughton, Ipswich, Suffolk IP8 3AW
Ingham, Mr R.	Kemin (UK) Ltd, Becor House, Green Lane, Lincoln, LN6 7DL
Jackson, Mr J.	Nutec Ltd, Eastern Avenue, Lichfield, Staffs
Jagger, Dr S.	Dalgety Agriculture Ltd, 180 Aztec West, Almondsbury, Bristol, BS12 4TH
Janes, Mr J.	Criddle Billington Feeds, Warrington Road, Glazebury, Warrington, Cheshire WA3 5PU
Jardine, Mr G.	Unitrition International Ltd, Barlby Road, Selby, N Yorks
Johnson, Miss S.	Kemin (UK) Ltd, Becor House, Green Lane, Lincoln, LN6 7DL

Jones, Mr H.	Heygate & Sons Ltd, Bugbrooke Mill, Bugbrooke, Northampton
Jones, Mr M.G.S.	Chad Associates, St Chad's Cottage, High Street, Pattingham, Wolverhampton WV6 7BQ
Jordan, Mr K.	96 Duncrue Street, Belfast, BT3 9AR
Keeling, Mrs S.	Nottingham University Press, Unit 2, Manor Farm, Thrumpton, Notts NG11 0AX
Kennedy, Mr D.A.	International Additives, Old Gorsey Lane, Wallasey, Merseyside
Ketelaar, Ir G.G.	Pricor BV, PO Box 51, 3420 DB Oudewater, Holland
Keys, Mr J.	32 Holbrook Road, Stratford upon Avon, Warks CV37 9DZ
Khan, Miss N.	Misset International, 7000 BX Doetinchem, The Netherlands
Kitchen, Dr D.I.	AF Plc, Kinross, New Hall Lane, Preston, Lancs PR1 5JX
Knock, Mr W.D.	MAFF, Room 256, Ergan House, 17 Smith Square, London SW1P 3JR
Knox, Miss L.	University of Nottingham, Sutton Bonington Campus, Loughborough, Leics LE12 5RD
Lamming, Prof G.E.	University of Nottingham, Sutton Bonington Campus, Loughborough, Leics LE12 5RD
Lane, Mr P.	Parnutt Foods Ltd, Hadley Rod, Woodbridge Industrial Estate, Sleaford, Lincs NG34 7EG
Lee, Dr P.	ADAS Rosemaund, Preston Wynne, Hereford HR1 3PQ
Lewis, Dr M.	Scottish Agricultural College, Bush Estate, Penicuik, Midlothian, EH26 0QE
Lima, Mr S.	Felleskjopet Rogaland Agder, 4001 Stavanger, Norway
Lindsay, Miss H.A.	Speciality Animal Feed Co Ltd, PO Box St 324, Southerton, Harare, Zimbabwe
Lister, Dr C.	Carrs Agriculture Ltd, Solway Mills, Silloth, CA5 4AJ
Longland, Dr A.C.	IGER, Plas Gogerdann, Aberystwyth, Dyfd, SY23 3EB
Lowe, Mr J.	Gilbertson & Page, PO Box 321, Welwyn Garden City, Herts

Lowe, Dr R.A.   Frank Wright Ltd, Ashbourne, Derbyshire,
                         DE6 1HA
Lucey, Mr P.   Dairy Gold Co-op Ltd, Lombardstown, Co
                         Cork
Lyons, Dr P.   Alltech Inc, Biotechnolgoy Center, 3031
                         Catnip Hill Pike, Nicholasville, Kentuck
                         40356 USA
MacDonald, Mr P.   David Moore (Favours) Ltd, 16/17 Abenbury
                         Way, Wrexham Ind Estate, Wrexham, Clwyd
Mackie, Mr I.L.   SCATS (Eastern Region), Robertsbridge,
                         East Sussex
Mafo, Mr A.   Proctors Ltd, Hi Peak Feeds Mill, 12
                         Ashbourne Road, Derbys
Malandra, Dr F.   Sildamin Spa, Sos Tegno di Spessa, 27010
                         Pavia, Italy
Marsden, Dr M.J.   Bibby Agriculture Ltd, Adderbury,
                         Oxfordshire, OX17 3HL
Marsden, Dr S.   Dalgety Agriculture Ltd, 180 Aztec West,
                         Almondsbury, Bristol BS12 4TH
Marsh, Mr S.P.   Rumenco, Stretton House, Derby Road,
                         Burton on Trent, Staffs DE13 0DW
Mather, Mr S.   Inroads International, 6 Post Office Court,
                         St Johns Street, Whitechurch, Salop
                         SY13 1QT
McAllan, Dr A.   IGER, Plas Gogerddan, Aberystwyth, SY23
                         3EB
McClelland, Mr D.J.   North Eastern Farmers Ltd, Bannermill,
                         Aberdeen, AB9 2QT
McGrane, Mr M.   Colborn Dawes Ltd, 25 Stockmans Way,
                         Belfast BT9 7JX
McIlmoyle, Dr W.A.   Nutrition Consultants, 2 Gregg Street,
                         Lisburn, Northern Ireland
McLean, Mr D.R.   WL Duffield & Sons Ltd, Saxlingham Thorpe
                         Mills, Norwich, Norfolk, NR15 1TY
Mills, Mr C.   University of Nottingham, Sutton Bonington
                         Campus, Loughborough, Leics LE12 5RD
Mitchell, Mr S.   BMS Computer Solutions Ltd, Sproughton
                         House, Sproughton, Ipswich, Suffolk IP8
                         3AW

Moore, Mr D.	David Moore (Flavours) Ltd, 16/17 Abenbury Way, Wrexham Ind Estate, Wrexham Clwyd
Morris, Mr M.	BOCM Pauls Ltd, 47 Key Street, Ipswich, IP4 1BX
Mounsey, Mr A.D.	HGM Publications, Abney House, Baslow, Derbyshire DE45 1RZ
Mounsey, Mr H.G.	HGM Publications, Abney House, Baslow, Derbyshire DE45 1RZ
Mounsey, Mr S.P.	HGM Publications, Abney House, Baslow, Derbyshire DE45 1RZ
Mul, Dr A.	BP Nutrition, Veerstraat 38, 5831 7N Boxmmer, The Netherlands
Munford, Dr A.G.	University of Exeter, Dept MSOR, North Park Road, Exeter EX4 4QE
Murray, Mr F.	Dairy Crest Ingredients, Philpot House, Rayleigh, Essex
Nelson, Ms J.	UKASTA, 3 Whitehall Court, London, SW1A 2EQ
Newbold, Dr J.R.	BOCM Pauls Ltd, PO Box 39, 47 Key Street, Ipswich, IP4 1BX
Newcombe, Mrs J.	University of Nottingham, Sutton Bonington Campus, Loughborough, Leics LE12 5RD
Nolan, Mr J.	Feed Flavours Inc, 265 Alice Street, Wheeling, Illinois, USA
O'Grady, Dr J.	IAWS Group Plc, 151 Thomas Street, Dublin 8
Offer, Dr N.W.	Scottish Agricultural College, Auchincruive, Ayr, KA6 5HW
Overend, Dr M.A.	Nutec Ltd, Eastern Avenue, Lichfield, Staffs
Owers, Dr M.	BOCM Pauls Ltd, PO Box 39, 47 Key Street, Ipswich, Suffolk IP4 1BX
Packington, Mr A.J.	Colborn Dawes Nutrition, Heanorgate, Heanor, Derbyshire
Partridge, Dr G.G.	Finnfeeds International Ltd, Market House, Ailesbury Court, High Street, Marlborough Wilts SN8 1AA
Pass, Mr R.T.	United Malt & Grain Distillers, Distillers House, 33 Ellersley Road, Edinburgh, EH12 6JW

Paul, Mr J.M.	BOCM Pauls Ltd, PO Box 39, 47 Key Street, Ipswich, Suffolk IP4 1BX
Payne, Mr J.D.	Borregaard Lignotech, 217 Kingsley Park, Whitchurch, Hants, RG28 7HA
Pearson, Mr A.	Hoechst Animal Health, Walton Manor, Walton, Milton Keynes, MK7 7AJ
Perrott, Mr G.	British Sugar, PO Box 11, Oundle Road, Peterborough, PE2 9QX
Perry, Mr F.	Rowett Research Services Ltd, Greenburn Road, Bucksburn, Aberdeen, AB2 9SB
Petersen, Miss S.T.	University of Nottingham, Sutton Bonington Campus, Loughborough, Leics LE12 5RD
Pettrigrew, Prof J.	University of Minnesota, Department of Animal Science, 1988 Fitch Avenue, St Paul, MN 55108 USA
Phillips, Mr G.	Silo Guard Europe, Greenway Farm, Charlton Kings, Cheltenham, Glos GL52 6PL
Pickford, Mr J.R.	Bocking Hall, Bocking Church Street, Braintree CM7 5JY
Pike, Dr I.H.	IFOMA, 2 College Yard, Lower Dagnall Street, St Albans, Hertfordshire AL3 4PE
Piva, Dr A.	Istituto di Zootecnia, Via Tolara do Sopra, 50 Ozzano-Bologna, Italy
Piva, Prof G.	Faculty of Agriculture, Via e Parmense 84, 29100 Piamense, Italy
Plowman, Mr G.B.	GW Plowman & Sons Ltd, Selby House, High Street, Spalding, Lincs PE11 1TW
Poornan, Mr P.	Lys Mill Ltd, Watlington, Oxon OX9 5ES
Povey, Dr G.M.	J Bibby Agriculture Ltd, Abberbury, Banbury, Oxon OX17 3QX
Putnam, Mr M.	Roche Products Ltd, PO Box 8, Welwyn Garden City, Herts, AL7 3AY
Rae, Dr R.C.	Premier Nutrition Products Ltd, The Levels, Rugeley, Staffs, WS15 1RD
Ramberg, Miss K,	Svenska Foder, Box 673, S-531 16 Lidkoping, Sweden
Raper, Mr G.J.	Laboratories Pancosma (UK) Ltd, Crompton Road Industrial Estate, Ilkeston, Derbyshire DE7 4BG

Record, Mr S.J.	Fishers Nutrition Ltd, Cranswick, Driffield, N Humberside
Redshaw, Mr M.	University of Nottingham, Sutton Bonington Campus, Loughborough, Leics LE12 5RD
Reeve, Dr A.	ICI Nutrition, Alexander House, Crown Gate, Runcorn, Cheshire WA7 2UP
Reeve, Mr J.R.	RS Feed Blocks, Orleigh Mill, Bideford, Devon
Retter, Dr W.	Heygate & Sons Ltd, Bugbrooke Mill, Bugbrooke, Northampton
Rice, Dr D.	Nutrition Services (Int) Ltd, 211 Castle Road, Randalstown, N Ireland
Rigg, Mr G.J.	Elanco Animal Health, Dextra Court, Chapel Hill, Basingstoke, Hampshire RG21 2SY
Roberts, Mr J.C.	Harper Adams College, Newport, Shropshire
Robertson, Mr S.	Hannah Research Institute, Ayr, KA6 5HL
Robinson, Mr D.K.	Favor Parker Ltd, The Hall, Stoke Ferry, Kings Lynn, PE33 9SE
Roele, Mr D.	Kemin Europa NV, Industrie Zone Wolfstee, 2410 Herentals, Belgium
Rose, Dr S.P.	Harper Adams Agric College, Newport, Shropshire, TF10 8NB
Rosen, Dr D.G.	Consultant, 66 Bathgate Road, London, SW19 5PH
Rosillo, Mr J.	12 Upminster Drive, Arnold, Nottingham
Ross, Mr F.	Lloyds Animal Feeds Ltd, Morton, Oswestry, Shropshire
Russell, Miss S.	Farmlab, Whestone Magna, Lutterworth Road, Whetstone, Leics LE8 6NB
Sanderson, Dr D.R.	IGER, Plas Gogerddan, Aberystwyt, Dyfed, SY23 3EB
Scribante, Dr P.	Laboratoires Pancosma SA, 6 Voie-des-Trax/cp 143, CH-1218 Grand-saconnex, Geneva, Switzerland
Shawcross, Mr C.	Anitox House, 80 Main Road, Earls Barton, NN6 0H5
Sheehy, Mr N.	Nutec, Greenhills Road, Tallaght, Dublin 24, Ireland
Shipton, Mr P.	Dardis & Dunns Coarse Feeds, Ashbourne, Co Meath

Shorrock, Dr C.	FSL Bells, Hartham, Corsham, Wilts SN13 0QB
Short, Miss F.	University of Nottingham, Sutton Bonington Campus, Loughborough, Leics LE12 5RD
Shrimpton, Dr D.	International Milling, Turret House, 171 High Street, Rickmansworth, Herts WD3 1SN
Shurlock, Dr T.G.H.	Lucta SA, PO Box 1112, 08080 Barcelona, Spain
Sigstadstoe, Mr H.	Felleskjopet Fortuvikling, Postboks 3771, Granaslia, N-7002 Trondheim, Norway
Silvester, Mr D.	Cyanamid UK, Fareham Road, Gosport, PO13 0AS
Silvester, Miss L.	Dalgety Agriculture Ltd, 180 Aztec West, Almondsbury, Bristol, BS12 4TH
Sissins, Mr	Bibby Agriculture Ltd, Adderbury, Banbury, Oxon OX14 1DJ
Sketcher, Mrs S.	Trouw Nutrition (UK) Ltd, Wincham, Northwich, Cheshire CW9 6DF
Sloan, Dr B.K.	Rhone Poulenc Animal Nutrition, 42 Aristide Briand Avenue, BP 100, 92164 Antony Cedex, France
Spencer, Mr P.G.	Bernard Matthews Plc, Gt Witchingham Hall, Norwich NE9 5QD
Stainsby, Mr A.K.	Bransby Agric Trading Assoc, Norton Road, Malton, North Yorks Y017 0NU
Stebbens, Dr H.R.	Crina (UK) Ltd, PO Box 111, New Malden, Surrey KT3 4YB
Steen, Dr R.W.J.	Agricultural Research Instit, Hillsborough, Co Down, N Ireland
Stockhill, Mr P.	Harboro Farm Sales Ltd, Howdenshire Way, Knedlington Road, Howden, Goole N. Humbs DN14 7HZ
Street, Mr C.	Berk Ltd, Priestley Road, Basingstoke, Hants, RG24 9QB
Sumner, Dr R.	Midland Shires Farmers Ltd, Defford Mill, Earls Croome
Taylor, Dr A.J.	Roche Products Ltd, Heanor Gate, Heanor, Derbyshire

Thompson, Mr D.	Rightfeeds Ltd, Castlegarde, Cappamore, Co Limerick
Thompson, Mr M.	Sheldon Jones Agriculture, Priory Mill, West Street, Wells, Somerset BA5 2HL
Thompson, Mr R.	AF Plc, Kinros, New Hall Lane, Preston, Lancs
Thorne, Dr C.	Waltham Centre for Pet Nutr., Freeby Lane, Waltham-on-the-Wolds, Melton Mowbray, Leics LE14 4RT
Tibble, Mr S.	SCA Nutrition Ltd, Maple Mill, Dalton Airfield Industrial Estate, Thirsk
Tice, Mr G.A.	Elanco Animal Health, Dextra Court, Chapel Hill, Basingstoke, Hampshire RG21 2SY
Tuck, Mr K.	Alltech Ireland Ltd, 28 Cookstown Industrial Estate, Tallaght, Dublin 24
Twigge, Mr J.	Trouw Nutrition (UK) Ltd, Wincham, Northwich, Cheshire, CW9 6DF
Unsworth, Dr E.F.	AESD, DANI, Newforge Lane, Belfast
Van Der Aar, Dr P.Y.	De Schothorst, PO Box 533, 8200 AM Lelystad, Holland
Vanstone, Mr M.	Crediton Milling Co Ltd, Fordton Milols, Crediton, Devon EX17 3DH
Vernon, Dr B.G.	BOCM Pauls Ltd, PO Box 339, 47 Key Street, Ipswich, IP4 1BX
Voragen, Prof A.F.G.	Wageningen Agric University, PO Box 8129, 6700 EV Wageningen, Holland
Vromant, Mr F.	Radar NV, Dorpsstraat 4, 9800 Deinze-Belgium
Wales, Mr C.	Volac Ltd, Orwell, Royston, Herts SG8 5QX
Walker, Mr A.W.	ADAS Gleadthorpe, Meden vale, Mansfield, Notts
Wallace, Mr J.R.	Nutrition Trading Ltd, Orchard House, Manor Drive, Morton Bagot, Studley Warks B80 7ED
Ward, Dr W.R.	University of Liverpool, Leahurst, Neston, South Wirral, L64 7TE
Wareham, Dr C.N.	Grain Harvesters Ltd, The Old Colliery, Wingham, Canterbury, Kent CT3 1LS
Waters, Dr C.J.	British Sugar, PO Box 11, Oundle Road, Peterborough PE2 9QX

Weeks, Mr R.H.	BOCM Pauls Ltd, 141 Brierley Road, Walton Summit, Preston PR5 8AH
White, Mr A.R.	Herd Care Consultancy, 4 Woodlands Meadow, Eaves Green, Chorley, Lancs PR7 3QH
Wilkinson, Dr R.G.	Harper Adams Agric. College, Newport, Shropshire, TF10 7ND
Williams, Dr D.	Anitox Ltd, Anitox House, 80 Main Road, Earls Barton, Northants NN6 0HJ
Winwood, Mr J.	Rhone-Poulenc Chemicals Ltd, Poleacre Lane, Woodley, Stockport
Wiseman, Dr J.	University of Nottingham, Sutton Bonington Campus, Loughborough, Leics LE12 5RD
Woolford, Dr M.	Alltech Ltd, 16117 Abenbury Way, Wrexham Industrial Estate, Wrexham, Clwyd LL13 9UZ
Wright, Mr I.D.	Microferm Ltd, Spring Lane North, Malvern Link, Worcs
Youdan, Dr J.	Nutrimix, Boundary Industrial Estate, Boundary Road, Lytham, Lancs FY8 5HU
Zwart, Mr J.E.M.	Hydro Agri Rotterdam BV, PO Box 58, 3130 AB Vlaardingen, Holland

INDEX